Strategic Studies Institute
and
U.S. Army War College Press

ASIA-PACIFIC:
A STRATEGIC ASSESSMENT

David Lai

May 2013

The views expressed in this report are those of the author and do not necessarily reflect the official policy or position of the Department of the Army, the Department of Defense, or the U.S. Government. Authors of Strategic Studies Institute (SSI) and U.S. Army War College (USAWC) Press publications enjoy full academic freedom, provided they do not disclose classified information, jeopardize operations security, or misrepresent official U.S. policy. Such academic freedom empowers them to offer new and sometimes controversial perspectives in the interest of furthering debate on key issues.

*****

FOREWORD

Dr. David Lai provides a timely assessment of the geostrategic significance of Asia-Pacific. His monograph is also a thought-provoking analysis of the U.S. strategic shift toward the region and its implications. Dr. Lai judiciously offers the following key points. First, Asia-Pacific, which covers China, Northeast Asia, and Southeast Asia, is a region with complex currents. On the one hand, there is an unabated region-wide drive for economic development that has been pushing Asia-Pacific forward for decades. On the other, this region is troubled with, aside from many other conflicts, unsettled maritime disputes that have the potential to trigger wars between and among Asia-Pacific nations.

Second, on top of these mixed currents, China and the United States compete intensely over a wide range of vital interests in this region. For better or for worse, the U.S.-China relationship is becoming a defining factor in the relations among the Asia-Pacific nations. It is complicating the prospects for peace and the risks of conflict in this region, conditioning the calculation of national policies among Asia-Pacific nations and, to a gradual extent, influencing the future of global international relations.

Third, the U.S. strategic shift toward Asia-Pacific is, as President Obama puts it, not a choice but a necessity. Although conflicts elsewhere, especially the ones in the Middle East, continue to draw U.S. attention and consume U.S. foreign policy resources, the United States is turning its attention to China and Asia-Pacific.

Fourth, in the mid-2000s, the United States and China made an unprecedented strategic goodwill exchange and agreed to blaze a new path out of the tragedy that often attends great power transition. It was a giant step in the right direction. However, it did not set U.S.-China relations forever. The two great powers can still encroach upon each other's core interests and overreact to each other's moves. In addition, the United States is also indirectly and deeply involved in many of China's disputes with its neighbors. These conflicts could lead China and the United States into unwanted wars.

Fifth, at this time of U.S. strategic reorientation and military rebalancing toward Asia-Pacific, the most dangerous consideration is that Asia-Pacific nations having disputes with China can misread U.S. strategic intentions and overplay the "U.S. card" to pursue their territorial interests and challenge China. China, on the other hand, believing that the United States intends to complicate China's relations with other Asian nations through actions on behalf of U.S. partners just short of shedding blood, could resort to strong and assertive actions to "silence" its opponents. The U.S. dilemma is how to maintain regional order in Asia-Pacific while not inadvertently encouraging China and its disputants to take reckless actions.

Finally, territorial dispute is becoming an urgent issue in the Asia-Pacific. China's dilemma centers on settling its territorial disputes. China appears to believe that time is not on its side—the longer it defers the issue, the stronger its opponents hold on the disputed territories becomes, further weakening China's position. There is ample evidence that while China still advocates shelving the disputes for the future, it is also making efforts to gain control of disputed territories. How China settles its disputes with its neighbors has become a very relevant issue, notwithstanding China's promise to settle these disputes peacefully and through bilateral consultations. With China blaming the United States for interfering and complicating the negotiations, can the United States and other Asia-Pacific nations give China the benefit of the doubt?

The Strategic Studies Institute and the U.S. Army War College Press is pleased to offer this monograph as a contribution to understanding the national security landscape of the Asia-Pacific of today and tomorrow. This analysis should be especially useful to U.S. strategic leaders as they seek to address the complicated interplay of factors and implications related to the U.S. strategic shift and military rebalancing toward Asia-Pacific.

DOUGLAS C. LOVELACE, JR.
Director
Strategic Studies Institute and
    U.S. Army War College Press

ABOUT THE AUTHOR

DAVID LAI is a Research Professor of Asian Security Affairs at the Strategic Studies Institute (SSI) and U.S. Army War College (USAWC) Press of the U.S. Army War College. Before joining SSI and USAWC Press, he was on the faculty of the U.S. Air War College. Having grown up in China, Dr. Lai witnessed China's "Cultural Revolution," its economic reform, and the changes in U.S-China relations over the years. In his research, Dr. Lai focuses on U.S.-Asian, especially U.S.-China, political and military affairs. He has also made significant contribution to the study of Chinese strategic thinking and operational art. Dr. Lai's most recent publication is the book, *The United States and China in Power Transition* (Carlisle, PA: Strategic Studies Institute, U.S. Army War College, 2011). Dr. Lai holds a bachelor's degree from China and a master's degree and a Ph.D. in political science from the University of Colorado.

SUMMARY

This analysis has four objectives: first, it puts the key trends of Asia-Pacific's geo-economic, political, and security affairs in perspective; second, it highlights the defining aspects in this region's complicated interstate relations; third, it points out the dilemmas confronting the key players in the region; and fourth, it draws attention to the most dangerous potential impacts of the region's outstanding conflicts.

Key points are as follows:

- Asia-Pacific, which covers China, Northeast Asia, and Southeast Asia, is a region with complex currents. On the one hand, there is an unabated region-wide drive for economic development that has been pushing Asia-Pacific forward for decades. On the other hand, this region is troubled with, aside from many other conflicts, unsettled maritime disputes that have the potential to trigger wars between and among the Asia-Pacific nations.

- On top of these mixed currents, there is an intense competition between China and the United States over a wide range of vital interests in this region. For better or for worse, the U.S.-China relationship is becoming a defining factor in the relations among Asia-Pacific nations. It is complicating the prospects for peace and the risks of conflict in this region, conditioning the calculation of national policies among Asia-Pacific nations and, to a gradual extent, influencing the future of global international relations.

- The U.S. strategic shift toward Asia-Pacific is, as President Barack Obama puts it, not a choice, but a necessity. Although conflicts elsewhere, especially the ones in the Middle East, continue to draw U.S. attention and consume U.S. foreign policy resources, the United States is turning its full attention to China and Asia-Pacific.

- In the mid-2000s, the United States and China made an unprecedented strategic goodwill exchange and agreed to blaze a new path out of the tragedy of great power transition. It was a giant step in the right direction. However, it does not take care of U.S.-China relations forever. These two great powers can still overstep the boundaries of each other's core interests and overreact to each other's moves. In addition, the United States is either indirectly or deeply involved in many of the disputes between China and its neighbors. These conflicts all run the risk of involving China and the United States in unwanted wars.

- As the United States makes its strategic shift and rebalances its military toward Asia-Pacific, it is faced with the problem that its Asia-Pacific allies who are pursuing territorial disputes with China will misread U.S. intentions and overplay the "U.S." card. On the other hand, if China believes that U.S. efforts are simply an attempt to complicate China's relations with its neighbors—without actually shedding any blood—it may take strong and assertive action to "silence" its opponents.

- For the United States, its dilemma is how to uphold the regional order in Asia-Pacific while not emboldening China and China's disputants to take reckless acts against each other.

- For China, its dilemma is when and how to settle its territorial disputes. It appears that China believes time is not on its side — the longer China defers the issue, the stronger its opponents' hold on the disputed territories, further weakening China's position. There is ample evidence that while China is still advocating shelving the disputes for the future, it is still making efforts to gain control of the disputed territories. Territorial dispute is becoming an urgent issue in Asia-Pacific. A more relevant question thus has become how China settles the disputes with its neighbors. China has promised to settle the disputes peacefully and through bilateral consultations. China also blames the United States for interfering and complicating the negotiations. Can the United States and the Asia-Pacific nations give China the benefit of the doubt?

There is no easy answer to these inextricable dilemmas. All in Asia-Pacific must walk a fine line in managing these conflicts.

ASIA-PACIFIC:
A STRATEGIC ASSESSMENT

Our Nation is at a moment of transition.

> President Barack Obama,
> January 2012

Accordingly, while the U.S. military will continue to contribute to security globally, we will of necessity rebalance toward the Asia-Pacific region.

> Department of Defense:
> *Sustaining U.S. Global Leadership:*
> *Priorities for 21st Century Defense,*
> January 2012

OVERALL ASSESSMENT

Asia-Pacific, which covers China, Northeast Asia, and Southeast Asia, is a region with complex currents. On the one hand, there is an unabated region-wide drive for economic development that has been pushing Asia-Pacific forward for decades. On the other hand, this region is troubled with, aside from many other conflicts, unsettled maritime disputes that have the potential to trigger wars between and among the Asia-Pacific nations. On top of these mixed currents, there is an intense competition between China and the United States over a wide range of vital interests in this region. For better or for worse, the U.S.-China relationship is becoming a defining factor in the relations among the Asia-Pacific nations. It is complicating the prospects for peace and the risks of conflict in this region, conditioning the calculation of national policies among the Asia-Pacific nations and, to a gradual extent, influencing the future of global international relations.

THE CHANGING GEOSTRATEGIC LANDSCAPE

Economy is a key driver of politics, domestic as well as international. The most striking characteristic of the Asia-Pacific is precisely its member nations' relentless pursuit of wealth and struggle for their economic development. Their efforts over the decades have led to a gradual shift of geostrategic power in the world and an increasing influence of this region as well.

As shown in Figure 1, Asia-Pacific today commands close to a quarter of the world's annual gross domestic product (GDP). By many measures, this global economic power distribution is expected to continue its tilt toward the Asia-Pacific region in the years ahead, turning the speculation of geostrategic shift of power center from the Atlantic to the Pacific into reality. Indeed, with infrastructures, including aggressive national economic development policies, well-facilitated centers of industrial production, spreading means of transportation, abundant supply of labor, operational and, more importantly, intellectual and engineering strategies well set in place, Asia-Pacific has been one of the most attractive destinations for foreign direct investment, trade, manufacture, and many other major business operations. Simply put, Asia-Pacific is poised to become the economic powerhouse of the unfolding Pacific Century.

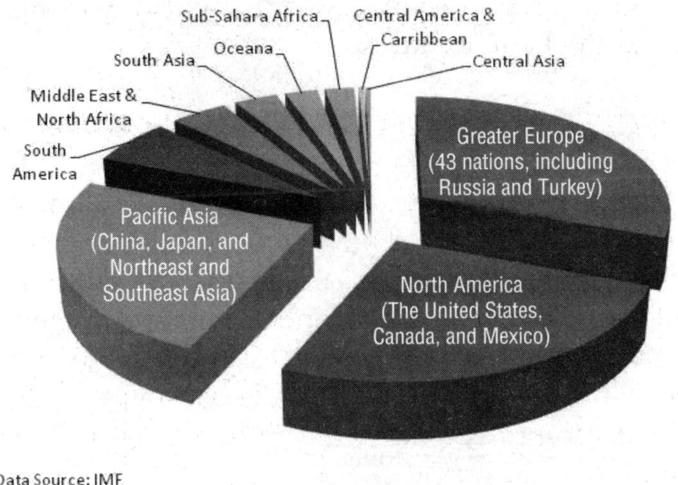

Figure 1. Distribution of GDP, 2011.

CHINA RISING

At the forefront of Asia-Pacific's economic development is the rising China. In 1978, China embarked on its modernization mission. At that time, few had expected China's efforts would create anything spectacular. After all, China had suffered through several false starts at modernization in the past (in the late Qing Dynasty following the invasions by foreign powers, during the Republic of China [ROC] and the People's Republic of China [PRC] under Chairman Mao's rule).[1] Yet by the late 1980s, China's move turned out to be genuine. A decade later, China's economic development (or China's rise, as it is more commonly known) started to take off. China's rapid rise is well captured in Figure 2. One can see that by the late 2000s, China surpassed Germany and Japan (and many other great powers along the way) to become the second largest economy in the world, trailing only behind the United States.

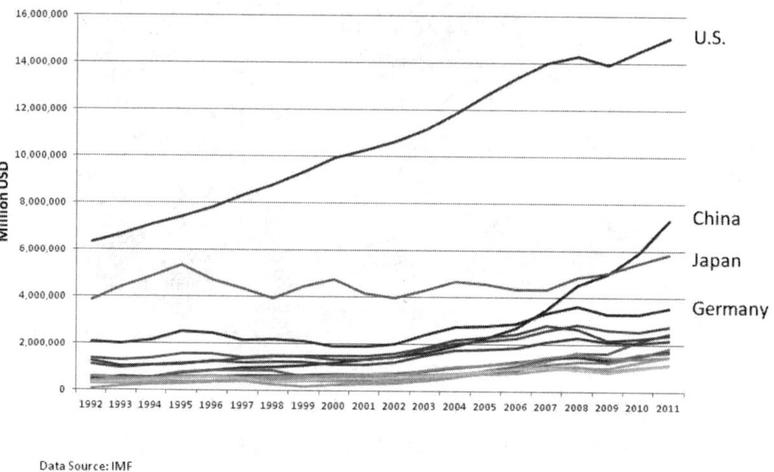

Figure 2. Top 15 Largest GDP Distribution, 1992-2011.

Thus, in a matter of 3 decades, China has gone from the brink of collapse (following decades of Mao-fabricated political movements, international isolation, economic starvation, and many other national malfunctions and disasters, the closest reflections of which can be found in today's North Korea) to the center stage of world economic development. It has been named by the United Nations (UN) as the world's new growth engine in recent years.

But the growth of its GDP is only one side of the China story.[2] China's per capita income level is still very low relative to many other nations, especially the other great powers. When its GDP is divided by the 1.4 billion Chinese people, China's per capita income ranks No. 96 among the 190-plus nations in the world (as of 2012).[3] Chinese leaders presumably have no illusion of where China stands among the nations. They have set a 30-year goal to bring China's standard of living, as measured by per capita GDP, into the ranks of the world's top 20 wealthiest nations.[4] This is a very ambitious plan. Yet by many measures, China has a good chance to bring about its dream.[5]

Geostrategic Impact of the Rising China.

China is not an ordinary nation. It is a great power by design (the most essential assets are geographic, demographic, and cultural). Once China develops, the external impact of its development will be extraordinary. (Recall Napoleon's remarks that when China awakens, it will shake the world).

The most significant external impact of China's rise is the consequential pressure it places on the onset of a power transition between China and the United States, the essence of which concerns both the current international system and the future of international relations.[6] The United States is the principal creator and caretaker of the extant international system. China, however, shared no part in the making of it following the end of World War II. Worse yet, the United States did not even acknowledge the founding of the new China for its initial 30 years (from 1949 to 1979), leaving China as a disgruntled outsider that sought the destruction of the U.S.-led international system.

When China becomes more powerful as a result of its internal economic development, there is concern that China will continue to press change to the U.S.-led international order and initiate a confrontation with the United States and its allies. There are several key reasons for this expectation. First, former rising great powers have all tried to alter the international order that they believed worked against their interests. Why should China be an exception? Second, China is controlled by a government that does not share fundamental values with the United States and its allies. A clash between China and the U.S.-led West over the basic principles of the extant international system may be an unavoidable course of action. The question is whether China's effort to promote change is civilized or violent, the latter of which is traditionally a recipe for war. Finally, China has the capacity and potential to become a superpower, possibly eclipsing the United States in the future. China also arguably has the ambition to make a new world in its image.[7] As the old saying goes, money makes the mare go; there is no telling what China can do with added national wealth on its rising path.

Considering all of the above and presumably more, the "China threat" is only a natural outcry in the United States and among its allies. It underscores the fact that the U.S.-China power transition is not an ordinary great power struggle, but a titanic shift of world power that has traditionally led to the change of world leadership in the political, economic, security, cultural, and other aspects of the international system.[8]

Changes in international relations have always come with a price. This is especially true with great power transitions. Indeed, throughout history, great powers have had to settle their differences on the battlefield and come to terms with the new realities of international relations with bloodshed.

Chinese and U.S. leaders have been informed of these tragedies.[9] The United States, in particular, has been concerned with the China threat ever since

China showed signs of economic-driven advances. China, in turn, is afraid that its rise can be derailed if it comes to a premature confrontation with the United States. Thus, in an attempt to counter the China threat outcry, China put forward an assertion of its peaceful development in 2003. The key components of this assertion are, first and foremost, China promises not to challenge U.S. supremacy (at least for now and presumably in a foreseeable future) on the condition that the United States does not step on China's core interests (i.e., national sovereignty and territorial integrity, but this condition, as seen in the sections that follow, is almost impossible); second, Chinese leaders point out that China no longer seeks the destruction of the present international order, but instead, has been making efforts to integrate itself into this system (with the exception of its government, of course); and third, China promises to avoid mistakes made by past great powers in similar power transition processes.[10] With these promises, China asked the United States to give its peaceful rise a chance for success.

Two years later, in 2005, the United States responded. China was clearly not disappointed. Deputy Secretary of State Robert B. Zoellick delivered the message. He first complimented China's progress over the previous 30 years. Then he reminded China of the values of the current U.S.-led international order, from which China has benefited tremendously. He subsequently asked China to become a responsible stakeholder of this system.[11]

Zoellick's remarks, in essence, acknowledged the fact that a rising China was not stoppable and that the United States would be better off trying to shape the direction of China's rise and manage the impact of this rising great power.[12] His call for China to become a responsible stakeholder indicated that the United States had made a strategic adjustment in its dealing with China—instead of keeping China outside, the United States welcomed China to join the "board of trustees" of the U.S.-led international system and asked China to support the United States from the "inside" of this system. This interaction was, by all measures, a goodwill exchange between China and the United States. It marked the first time in the history of international relations that two great powers in power transition openly addressed the key issues involved in this process and promised to blaze a new path out of the deadly contest.[13]

While this is certainly a welcome step in the right direction, the U.S.-China goodwill exchange is nevertheless overshadowed by some contentious and almost intractable conflicts between the two nations. This brings us to another significant external impact of China's rise: China's unavoidable conflict with the United States in China's pursuit of its expanded national interests. When a great power becomes more powerful internally, there is a natural tendency for it to consolidate its interests externally. This maxim is particularly pertinent to China. Indeed, although China is one of the world's oldest and most continuous civilizations, it is a young nation with a great deal of unsettled nation-building business. The most outstanding are matters that Chinese leaders consider core interests of national unity and territorial integrity. As shown in Figure 3, many of China's disputes are in the Western Pacific (the ones in China's west, namely, unsettled border disputes with India and unsettling issues in Tibet and the Uyghur area are troublesome, but China has been able to hold those issues under control). At the top on the list is the Taiwan issue, the fate of which has been a point of contention between China and the United States for well over 60 years. In recent years, China has also had intense conflict with the United States over U.S. military activities in the Chinese-claimed maritime Exclusive Economic Zone (EEZ). Farther out in the East and South China Seas, China has territorial and EEZ disputes with Japan, the Philippines, Vietnam, Malaysia, Indonesia, and Brunei, respectively. The United States is involved indirectly but increasingly directly in China's disputes with its maritime neighbors.

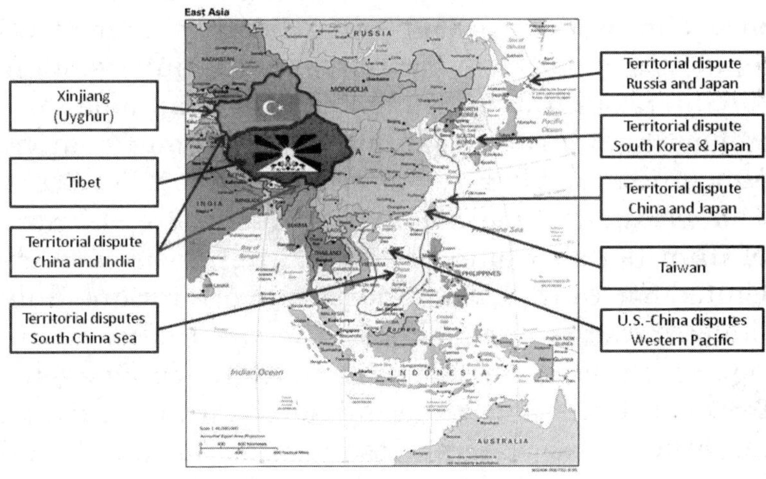

Figure 3. China's Disputes.

From Figure 3, one can also see that China is at the center of a ring of conflicts. These China-centered conflicts are another defining factor in Asia. China has made it clear that until it settles all of these conflicts, presumably in its favor, China will not be a full-fledged great power. But the challenging aspect of China's mission is that the United States has been involved directly and indirectly in all of the conflicts for decades. China must settle its relations with the United States every step of the way to accomplish its mission.

This is no doubt a huge undertaking. Unfortunately, it is by no means easy, for none of these conflicts has an attainable solution in sight. Worse, they are increasingly being complicated by the U.S.-China power transition—every Chinese move is being perceived as part of China's challenge to the United States. Chinese leaders, on the other side of the Pacific, believe that the United States has ill will toward China and makes trouble in all the disputes China has with its neighbors. Thus in dealing with those disputes, China feels compelled to prepare for possible U.S. intervention. A natural outgrowth of this "Chinese paranoia" is a military buildup overall, and more pointedly along the Chinese side of the Western Pacific.

The single instrumental impetus for China's military buildup is the Taiwan Strait crisis of 1995-96. At that time, Chinese leaders were furious with Taiwan's push for statehood. They sent harsh warnings to Taiwan, staged military exercises along the Taiwan Strait, and fired missiles toward Taiwan, splashing into waters close to the northern and southern tips of the island. These hostile exchanges prompted a U.S. military intervention with two U.S. aircraft carrier battle groups scrambling toward the Taiwan Strait.

Although the U.S. Naval forces moved cautiously around the troubled waters during the crisis, they nevertheless sent an unmistakable message to the Chinese, as in the words of the then Secretary of Defense William Perry, "Beijing should know—and this [the reinforced U.S. fleet] will remind them—that, while they are a great military power, that the premier, the strongest military power, in the Western Pacific is the United States."[14]

This U.S. reminder was certainly a bitter pill for the Chinese to swallow. Unfortunately, it was also one that electrified the Chinese political and military leaders. They subsequently undertook extraordinary measures to develop what we know today as the anti-access and area-denial (A2/AD) capabilities along China's maritime fronts in the Western Pacific (see Figure 4).[15]

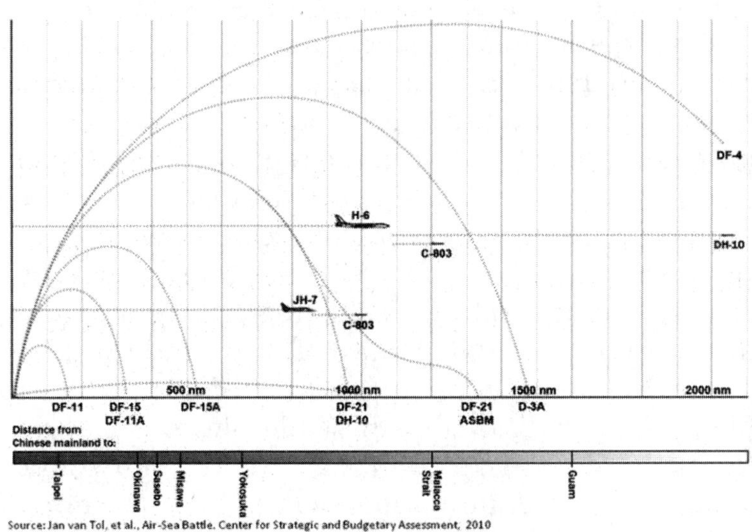

Source: Jan van Tol, et al., Air-Sea Battle. Center for Strategic and Budgetary Assessment, 2010

Figure 4. China and A2/AD.

China's A2/AD capabilities were initially developed to deal with a future Taiwan Strait crisis and possible U.S. military intervention. As these capabilities gradually come into operation, China can also employ them to deal with its "U.S. problem" at large in the Western Pacific.

U.S. STRATEGIC SHIFT

The United States has been trying to respond to the rising China issue since the George H. W. Bush administration in the early 1990s. However, "burning issues" elsewhere have kept the United States busy in other parts of the world (the wars in Iraq and Afghanistan, for instance) and unable to develop a coherent response to China's monumental challenge until the Obama administration took office in 2009.

The Obama administration's move is now known as the U.S. strategic shift toward Asia-Pacific. The key elements are as follows: the United States will reassert its leadership in Asia-Pacific, regain its economic preeminence, continue to promote democracy, and reinforce the security order in the region. The execution of this strategic shift started with the Obama administration's "Returning to Asia parades" (a swing of the U.S. policy from the George W. Bush administration's alleged "benign neglect" of this region). Upon taking office, Secretary of State Hillary Clinton and President Obama made unprecedented first official visits to Asia-Pacific (traditionally, these high-level visits were first made to the European allies) with high sound-bite calls of "the United States is back" and "America's first Pacific President is here to lead."

Secretary Hillary Clinton characterized the U.S. effort as an action moving along six key lines:

> strengthening our bilateral security alliances; deepening our working relationships with emerging powers; engaging with regional multilateral institutions; expanding trade and investment; forging a broad-based military presence; and advancing democracy and human rights.[16]

Following these guidelines, the United States soon joined the Treaty of Amity and Cooperation in Southeast Asia. This act paved the way for Secretary Clinton and President Obama to attend the East Asia Summits in 2010 and 2011. In a strikingly different way, the United States outplayed China, which has been carefully following a low-key approach to deal with the Asia-Pacific nations in this high-level diplomatic arena.

Then the United States joined the Trans-Pacific Partnership (TPP). This act is widely seen as a U.S. move to use the TPP as a vehicle to promote a U.S.-led free trade zone in the Asia-Pacific. The United States had been the largest trading partner to all the Asian nations until the early 2000s, when China took over the crown. As shown in Figure 5, China's trade with the Asia-Pacific nations has steadily increased over the years. By 2010, the volume of China's trade with the Asia-Pacific nations was almost twice the size of that of the United States.

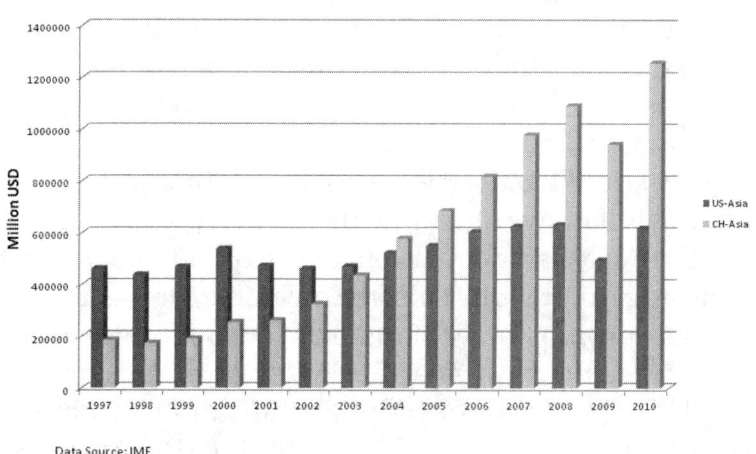

Figure 5. U.S.-Asia and China-Asia Trade, 1997-2010.

A recent report by the Associated Press points out the changing status of China and the United States as trading nations at a global scale:

> As recent as 2006, the U.S. was the larger trading partner for 127 countries, versus just 70 for China. By last year [2011] the two had clearly traded places: 124 countries for China, 76 for the U.S. . . . The findings show how fast China has ascended to challenge America's century-old status as the globe's dominant trader, a change that is gradually translating into political influence. They highlight how pervasive China's impact has been, spreading from neighboring Asia to Africa and now emerging in Latin America, the traditional U.S. backyard.[17]

Trade flows are good reflections of international relations. It is a convention that the more a nation trades with others, the closer its relations with the others will become. These mutually dependent relations also tend to change their policy preference and calculation toward one another. China's influence on Asia-Pacific nations has been on the rise accordingly. It is notable that these changes are also affecting the two strongest allies of the United States, Japan, and Korea, whose largest trading partner is no longer the United States, but China.

The United States is determined to reverse this trend. The TPP would allow the United States to break into the Asia-Pacific markets and expand U.S. export to this region. It would also be a vehicle for the United States to promote new standards for 21st century free trade agreements. It is a U.S. effort to jump-start the long-halted process for trade liberalization within the Asia-Pacific Economic Cooperation (APEC) framework. While the TPP will bring economic interests to the United States, the real U.S. intent is to counterbalance China, reduce China's influence in the Association of Southeast Asian Nations (ASEAN)-China Free Trade Area (ACFTA) and other China-led or China-involved regimes in East and Southeast Asia, and establish rules and codes of conduct for China to follow.[18] Of note is that China is not included in the TPP. There is already an expected uneasy reaction from China on this U.S. initiative.

In the meantime, the United States has also taken measures to reinforce the security order in Asia-Pacific. These measures are very straightforward. The United States first strengthens relations with its current allies and then tries to recruit new partners. There is certainly no lack of candidates for the United States. In the face of rapid Chinese economic expansion in Asia-Pacific and growing Chinese influence in many aspects of the region's relations, many Asia-Pacific nations want to develop relations with the United States and keep the United States as a provider of security in the region. This has long been a balancing act of the Asia-Pacific nations, although many of them try not to let the partnership with the United States become an overt U.S. leverage against China.

For its part, the United States understands that it is ultimately the one to bear the cost of providing the public good of common security in Asia-Pacific. The U.S. military has an indispensible responsibility in this regard. Thus, in January 2012, when President Obama rolled out a roadmap for the rebalancing of the U.S. military and its priorities in the years ahead, his strategic guidance states that, while the U.S. military will continue to maintain its global commitments, it will make a strategic shift to the Asia-Pacific region. Specifically, the United States will keep about

60 percent of its armed forces in the Asia-Pacific. In addition, the U.S. military will follow an Air-Sea Battle concept to develop capabilities to deal with China's A2/AD challenges:

> The United States must maintain its ability to project power in areas in which our access and freedom to operate are challenged... *Accordingly, the U.S. military will invest as required to ensure its ability to operate effectively in anti-access and area denial (A2/AD) environments* (emphasis original).[19]

Through these aggressive moves, the Obama administration puts U.S. diplomatic, information, military, and economic instruments (DIME) into full play. On many occasions, Secretary Clinton repeats the remarks that:

> From the very beginning, the Obama administration embraced the importance of the Asia-Pacific region. So many global trends point to Asia. It's home to nearly half the world's population, it boasts several of the largest and fastest-growing economies and some of the world's busiest ports and shipping lanes, and it also presents consequential challenges such as military buildups, concerns about the proliferation of nuclear weapons, natural disasters, and the world's worst levels of greenhouse gas emissions. It is becoming increasingly clear that in the 21st century, the world's strategic and economic center of gravity will be the Asia-Pacific, from the Indian subcontinent to the western shores of the Americas. And one of the most important tasks of American statecraft over the next decades will be to lock in a substantially increased investment — diplomatic, economic, strategic, and otherwise — in this region.[20]

Secretary Clinton also puts it categorically that "the future of politics will be decided in Asia, not Afghanistan or Iraq, and the United States will be right at the center of the action;" and the unfolding Pacific Century will be "America's Pacific Century."[21]

U.S.-CHINA CONTESTS IN ASIA-PACIFIC

While the U.S.-China power transition will take time to become truly global, it has already cast a big shadow on Asia-Pacific relations. Most, if not all, of the old problems and conflicts in this region between China, the United States, and China's neighbors have now taken on new significance. Many of the new developments in those conflicts will make sense only when they are put in the context of the U.S.-China power transition.

The Inextricable Dilemmas.

The discussion in this section tries to put the key conflicts in perspective. But before delving into the details, a highlight of the dilemmas confronting China, the United States, and the Asia-Pacific nations is in order.

China.

For China, it is the issue of timing and methods in handling its disputes with other nations. With respect to timing, Chinese leaders wish they would not have to come to a showdown with the other disputants in the next 30 years, so they will have time to turn China into a true great power. The obvious reason is that China needs a war-free environment to develop (conceivably not because the Chinese are inherently peace loving, as they have always claimed). China's concern is that premature confrontation would interrupt or even derail its mission. There is also an unspoken yet undeniable reason that 30 years down the road, China would be much more powerful, hence in a stronger position to settle the disputes in its favor.

However, with respect to the disputed territories, time is not on China's side. China does not have effective control over most of the disputed territories. In international practices, the nation that has effective control over its disputed territory has a better chance to win the case. That is certainly an established rule in the International Court of Justice (ICJ).[22] Thus, the longer the time passes by, the stronger the hold other disputants will have on those territories; and it would be more difficult for China to win the fight. Moreover, the other disputants may not answer China's call to shelve the disputes, but instead will take the time to alter the status quo. They will not sit idle and watch China becoming more powerful in the years to come. They will most likely take the time to develop their own defense capabilities and shore up outside support. All in all, China does not have the luxury to wait 30 years to settle these disputes.

With respect to the ways of dispute settlement, China also has a difficult choice to make. China has promised to make a peaceful rise. If it were to use force to settle its disputes, China would find its acts indefensible. China is also concerned that an armed conflict with the other disputants will provide testimony to the outcry that China is a threat and force the other disputants to gang up against China. However, there is ample evidence that the Chinese are trying to find ways to circumvent this dilemma. In recent years, there are mounting calls in China for the Chinese government to take stonger stands and use force if necessary to settle its disputes sooner rather than later. In Chinese characterizations, labor pains are easier to handle than growing pains. Time can easily wash away the memories of those short-term blows.

United States.

For the United States, the current situation is a matter of balancing its relations with China and its support of China's disputants. Doing too little runs the risk of emboldening China to take tougher stands against its neighbors. But doing too much would antagonize China and encourage China's disputants to overplay the "U.S. card," bringing the United States into unwanted fights with China. Both are dangerous aspects of the U.S. strategic shift and military rebalancing in this region. The United States is involved in many international conflicts. "Doing the right thing (strategically)" and "doing the things right (operationally)" are difficult choices for the United States everywhere and more so in Asia-Pacific because it involves a rising power, China, with many entangled interests and conflicts.

Asia-Pacific Nations.

For the Asia-Pacific nations, especially China's disputants, it is a matter of getting the best out of the U.S.-China competition. Their difficult choice is how to play the "U.S. card" to advance their interests and strengthen their position on the contested territories with China while not choosing sides between China and the United States over the two great powers' contested issues.

There is obviously no easy answer to any of the above-mentioned dilemmas. All have to walk a fine line in managing these conflicts.

The China-Taiwan-U.S. Issue.

The fight for the fate of Taiwan is a direct conflict between the United States and China. It has been a sticky issue between the two nations for well over 6 decades.

A Multi-Direction Tug of War.

China has an avowed mission to reunite with Taiwan. It has promised to use force to achieve this goal if peaceful means fail. The United States has a Taiwan Relations Act to ensure that no use of force is allowed to change the status of the island nation. Taiwan, however, is a complicated story. It has gone through changes from an agrarian society to an industrial powerhouse and from authoritarian rule to democracy. Since the mid 1990s, Taiwan has been subject to a constant internal tug of war between the two major political parties on the issue of pressing for Taiwan independence/statehood or maintaining its status quo, with the Democratic Progressive Party (DPP) pushing for the former, and the Kuomintang (KMT) upholding the latter.

The internal and external tug of wars triggered a Taiwan Strait crisis in 1995-96. In many ways, this crisis is a watershed event in the China-Taiwan-U.S. relations. First, the issue of Taiwan independence/ statehood came to the surface and has since become an openly contested and dividing issue in Taiwan's political life, especially during election times. Second, the United States made the first military intervention (though symbolically) as required by the Taiwan Relations Act. Finally, China started a military buildup to deal with both a possible future drive for Taiwan independence and a U.S. military intervention.

Swing of the Pendulum.

The year 2000 witnessed another landmark change in this multidirectional tug of war. The DPP won the presidential election and ended the KMT's 50-year rule of Taiwan. The core members of the DPP used to be political dissidents; some were persecuted by the KMT and others were underground activists. Most of them are advocates of Taiwan independence. With its control of the government, and under the leadership of a strongly pro-Taiwan independence president, the DPP was able to make an all-out campaign to promote its cause. In 2004, the DPP won a second term in office. It subsequently took more radical measures to advance the Taiwan independence agenda. Thus in the 8 years under its administration, the DPP pushed the Taiwan independence/statehood movement to the extremes and brought the China-Taiwan-U.S. relations to the brink of war several times.

China responded furiously to the DPP's moves. It adopted an Anti-Secession Law in 2005 in which China put down the conditions under which it would use force against Taiwan:

> In the event that the "Taiwan independence" seces-
> sionist forces should act under any name or by any
> means to cause the fact of Taiwan's secession from
> China, or that major incidents entailing Taiwan's se-
> cession from China should occur, or that possibilities

for a peaceful reunification should be completely exhausted, the state shall employ non-peaceful means and other necessary measures to protect China's sovereignty and territorial integrity.[23]

The United States worried that the DPP's reckless acts could provoke a war across the Taiwan Strait and force the United States to undertake military intervention again. In 2003, President George W. Bush put forward a timely warning to the leaders on the two sides of the Taiwan Strait that neither should take unilateral action to alter the status of Taiwan. In a pointed way, the Bush administration warned Taiwan's pro-independence leaders that they should not provoke a war with China by recklessly pressing their agenda, or they would have to bear the responsibility by themselves. At the same time, the Bush administration also put the Chinese leaders on notice that they should not use force to coerce unification, or the United States would intervene.[24] By removing the "strategic ambiguity" in the U.S. position that had confused Taiwan and China for years, the United States helped stabilize the situation in the Taiwan Strait.[25]

Return of the Pendulum.

The DPP-driven tensions came to a halt in 2008, with the KMT regaining control of the Taiwan government by a victory in both presidential and legislative elections. The KMT stands for eventual unification with China but insists that the unification should take place under democratic principles and, more specifically, when Chinese government becomes a democratic one. Since democratic government in mainland China is nowhere in sight, the KMT therefore makes it clear that there will be no rush for unification on

Taiwan's part. But in an attempt to ease China's concern, the KMT also promised China that there would be no push for Taiwan independence under its watch. In return, the KMT asked China to promise that there would be no use or threat of force at the same time. Hence came the KMT's, "Three No's" policy—no unification, no Taiwan independence, and no Chinese use of force."

China responded positively to the KMT's call. In the past 4 years, the Chinese government and the KMT administration joined hands to open direct air and sea travel routes and signed an Economic Cooperation Framework Agreement in 2010, a preferential trade agreement between China and Taiwan aimed at reducing tariffs and commercial barriers. These were very significant developments, since the two sides split after the Chinese Civil War in 1949. In 2012, the KMT won a second term in office. It appears that the KMT has gained endorsement from Taiwan's voters that its Three-No's policy and engagement with mainland China should be continued.

Chinese Concerns with the Impact of U.S. Strategic Shift on Taiwan.

However, with the U.S. strategic shift toward Asia-Pacific on the move, China is concerned that the DPP will try to take advantage of the U.S. move to rock the boat again. China's apprehension is manifold. In the political sense, the Chinese worry that the DPP may misread the U.S. intent to promote democracy and press the Taiwan independence agenda again. Chinese leaders believe that deep in their heart, the DPP leaders want Taiwan independence; and the DPP places its hope of success on: 1) the collapse of the mainland

Chinese government through democratic change or whatever reason; and, 2) U.S. support for Taiwan independence (or tacit connivance with the movement). Chinese leaders also believe that in previous confrontations, pro-Taiwan independence activists pushed the agenda, either because of U.S. behind-the-scene support or their misinterpretation of the U.S. strategic interests.

On the economic front, China is concerned that Taiwan may push for membership in the U.S.-led TPP. Membership in the TPP would help Taiwan gain international status (the DPP and KMT share common interest in this regard). But it would also ensure a fight between China and Taiwan, because China is concerned that Taiwan's membership in the TPP will provide Taiwan with another support for its quest for statehood.

In military affairs, China is afraid that the DPP may make use of the U.S. military's rebalancing toward Asia-Pacific and push for closer military cooperation with the United States. Pressing for more arms sales to Taiwan will be a sticky issue in this respect. Provision of arms of a defensive nature to Taiwan is part of the U.S. commitment to the defense of Taiwan in the Taiwan Relations Act (TRA, U.S. Public Law 96-8). Periodic U.S. authorization of arms sales has been a major point of contention between China and the United States and brought setbacks in U.S.-China relations in general and military-to-military exchanges in particular since the TRA came into effect in 1979.

There are deep political and cultural differences between the United States and China on this issue. To successive U.S. administrations, providing weapons to Taiwan is only meeting the demands from Taiwan and complying with the requirement of the TRA. In addition, as a gun-cultured nation, the United States acutely believes in gun-related self defense. Providing arms to Taiwan goes along with the prevailing American view that Taiwan is entitled to arm itself against threats from mainland China. Thus the question for the United States is not whether or not to provide arms to Taiwan, but how to conduct the business without the contentious repercussions from China. China, however, insists that the United States should not conduct this business in the first place. It wants the United States to stop the arms sales altogether. For the Chinese, it does not matter how the United States handles the business. As a result, U.S.-China dialogue on arms sales to Taiwan has never gone beyond these irreconcilable quarrels over the years.

Unfortunately, the contention is not going to go away any time soon. The United States will not do so and China does not have the power in the foreseeable future to force an end to this business, hawkish Chinese calls for a showdown notwithstanding.[26] Nevertheless, U.S. arms sales to Taiwan will force the two great powers to test each other's will time and again.

In sum, while there is a constructive trend in the China-Taiwan-U.S. relations, there are also risks for conflict. It is a case that must be handled with high-level attention.[27]

U.S.-China Conflict over the Exclusive Economic Zone.

This is another direct confrontation between China and the United States. It centers on the U.S. military activities in the Chinese-claimed EEZ. The issue stems from the two sides' diametrically-opposing views on the legal and practical nature of the U.S. military activities in this area. The opposing views and acts have gotten the two nations to confront each other in hostile ways in the Western Pacific. The U.S.-China power transition factor has only made this issue all the more contentious. Indeed, the U.S. call for the freedom of navigation, its denunciation of China's A2/AD strategy, and the development of the Air-Sea Battle plan are all parts of the U.S. countermeasures against the expanding Chinese reach in the Western Pacific.

U.S.-China Conflicts in Acts and Words.

The most notable confrontation so far is the collision of a U.S. EP-3 surveillance plane with a PLA fighter jet about 70 miles off China's southern coast over the South China Sea on April 1, 2001 (unfortunately, it was not an April Fool's Day joke).[28] Since then, China and the United States have continued to clash in the South and East China Seas. China has reportedly "harassed" the entire U.S. ocean surveillance fleet on various occasions such as the United States Naval Ship (USNS) *Bowditch* (September 2002), *Bruce C. Heezen* (2003), *Victorious* (2003, 2004), *Effective* (2004), *John McDonnell* (2005), *Mary Sears* (2005), *Loyal* (2005), and *Impeccable* (2009).[29] The latest incident in the South China Sea, involving the USNS *Impeccable*, generated a new round of outcry between the two nations. Dennis Blair, former Commander of the U.S. Pacific Command, called it the most serious confrontation since the EP-3 incident.[30]

In addition to the clashes over the surveillance ships, China has also taken issue with the U.S. aircraft carrier group conducting military exercises in the Yellow Sea and its occasionally transiting the Taiwan Strait.[31] China argues that the UN Convention on the Law of the Sea (UNCLOS) has established the

200-nautical mile (nm) EEZ between territorial waters and the high seas as a special area different from either and governed by its own rules. China also holds that freedom of navigation and over-flight in the EEZ have certain restrictions, namely, the activities must be peaceful and nonthreatening to the coastal nations. China charges that U.S. military surveillance ships and reconnaissance flights in the Chinese-claimed EEZ have hostile intent on China; and their actions therefore do not fall in the scope of peaceful and innocent passages. China has repeatedly asked the United States to reduce this activity and eventually put a stop to it.[32] The United States categorically rejects China's claims, insisting that China misinterprets the UNCLOS at best, but more pointedly, intentionally stretches the interpretation to stage this confrontation with the United States at worst. The United States holds that the UNCLOS sanctions on foreign military activities only include 12-nm territorial waters, but not the EEZ. The United States also argues that China's reservation to the UNCLOS on foreign military activities in the EEZ does not enjoy broad support from the other signatory parties. Indeed, of the 161 nations ratifying the treaty, only 14 reserve the right to require approval for foreign military activities in their claimed EEZ's. China's position therefore is an exception rather than the rule.[33]

The U.S. side also points to its experience with the Soviet Union during the Cold War. The two superpowers had an agreement to avoid incident on the high seas (the 1972 U.S.-Soviet Union Incidents-at-Sea Agreement [INCSEA]). When Soviet military vessels came to U.S. shores, the United States shadowed and watched them closely, but did not demand their departure as the Chinese do now. The Chinese argument that the United States would not tolerate Chinese military surveillance ships to get to U.S. shores does not stand. China has gone too far in its demands.[34]

Moreover, the U.S. side has also hinted to China that, as its interests expand globally, China may need to send its navy to faraway areas to protect those interests; it would be therefore in China's interest to keep the EEZ open for foreign military activities. China at this point does not buy these U.S. arguments. One example is that in its effort to fight against piracy at the Gulf of Aden, China explicitly asked for permission from the Somalia government to let the Chinese naval forces operate in the Somalia troubled waters. Aside from the above arguments, the United States holds that it has been conducting these businesses for well over 60 years, and no one is to tell the U.S. military to stop exercising its freedom of navigation in this part of the ocean.

On a more controversial note, Admiral Timothy Keating, while visiting Beijing as commander of the U.S. Pacific forces, put it on record that the United States does not need permission from China to sail its aircraft carrier group through the Taiwan Strait.[35]

National Power as the Final Arbiter?

These opposing views and confrontations are difficult to reconcile given that: 1) the two nations have had troubled relations throughout the years and do not trust each other; and, 2) the two are engaged in the ongoing power transition and the fight over the EEZ is a test case on the national strength of the two power transition contestants. The stakes are high on both sides.

Unfortunately, there is no easy fix to this problem. The UNCLOS is not likely to come up with a solution to these disputes any time soon. In the meantime, China and the United States have to take their arguments in their own hands to confront each other.

Unfortunately, with the absence of mutually (and internationally) acceptable grounds, the ultimate arbiter over the U.S. military activities in China's claimed EEZ will be the two nations' national power, especially their military power. China at this point, and for some time to come, does not have the capability to carry out its demands. It can only make repeated protests or harass the U.S. operation in the Chinese-claimed EEZ. However, China is making steady effort to improve its fighting capabilities. China's Marine Administration now has many well-equipped patrol ships and airplanes to do "law enforcement" acts in its claimed EEZ on a regular basis. China's navy will come to back up these acts if they encounter hostile acts from opponents. It will only be a matter of time before China will take a more forceful stand on this issue. According to the Pentagon's and other credible institutions' assessments, it will probably take China another 10 to 15 years to reach that level.[36] In the realm of international relations, this is a very short time span. Indeed, China has already made the call that the United States should prepare itself to accept this change and accommodate China's demands.

U.S.-China competition or confrontation in the Western Pacific is a difficult issue. It is unlike the two nations' encounters in other regions of the world, where there is a good chance that UN sanctions are in order or conflict is over less important interests. In these regions, the two nations could find it easier and more beneficial to cooperate. A prime example of this is the Chinese PLA Navy's escort mission in the Gulf of Aden and its cooperation with the United States and international forces. The two nations' fight in the Western Pacific is direct, with vital interests at stake; difficult for the small Asian nations to intervene; and difficult to settle the issues peacefully. It is an emerging reality that is difficult for the two nations to come to terms with in the context of their ongoing power transition. The U.S. rebalancing toward Asia-Pacific will likely put more pressure on this contested issue.

GREAT POWER COMPETITION IN NORTHEAST ASIA

Northeast Asia is a place where five of the world's most powerful nations meet: China, Japan, South Korea, Russia, and the United States (the only odd one out is North Korea). Three of them are the world's largest economies (the United States, China, and Japan) and largest militaries (China, the United States, and Russia). In political terms, the United States, Japan, and South Korea are champions of democracy; China is the largest authoritarian nation; and Russia is a bizarre mix of half-baked democracy and half-revived authoritarian rule.

The five great powers are strange bedfellows. The United States, Japan, and South Korea are related through democratic values and military alliances. China and Russia are strategic partners of convenience. All five great powers have been enemy to one another in the past. Although at times common interest dictates that they cooperate, the five powers nevertheless follow their own national interests to pursue their goals.

Conflict of interest is natural, but compromise of national interest is difficult. A prime example is the five powers' "romance" with North Korea in the past decade. Although their goals were the same—trying to prevent North Korea from developing nuclear weapons—they nonetheless brought different interests to the on-and-off Six-Party Talks and ended up accomplishing nothing.

The North Korea problem, ironically, was a good thing for the five great powers. After all, it offered them a common problem and forced them to cooperate to a good extent. But recently (since early 2010), a long-time divisive issue among the five powers has resurfaced to drive them apart. It is the issue of Japan's maritime territorial disputes with Russia, South Korea, and China. The United States was involved in the making of those disputes and has had a stake in the quarrels the whole time.

These disputes have almost intractable historical claims and contemporary circumstances, but none has a fair or attainable resolution in sight. Recent flare-ups have only further complicated the relations among the five great powers.

The Russo-Japanese Territorial Dispute.

The dispute is over the four islands in the southern tip of Russia's Kuril island chain, or the Northern Territories in Japanese terms. The islands are only a few miles off Japan's northern prefectures of Hokkaido (see the circled area in Figure 6). Russia took hold of those islands at the end of World War II as spoils of the great power post-war settlement and has maintained effective control ever since. Japan, however, claims sovereignty over those islands and has pressed for their return all along. This dispute has been a major obstacle in Japan's relations with the Soviet Union and its successor, Russia.

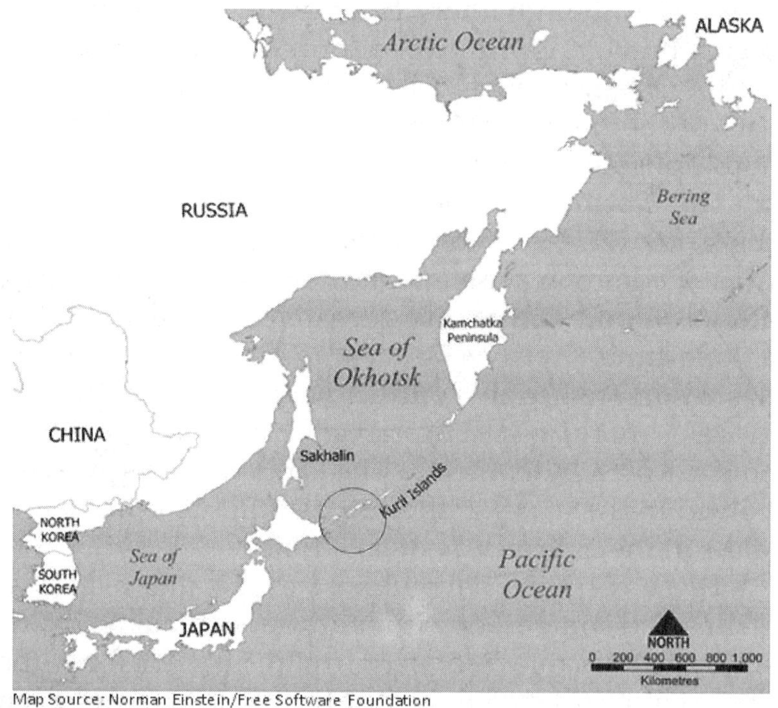

Map Source: Norman Einstein/Free Software Foundation

Figure 6. Russo-Japanese Territorial Dispute.

The United States supports Japan's claim over the islands and the efforts for their return, but declines to extend the U.S.-Japan mutual defense treaty coverage over those islands on the excuse that they are not under Japan's effective administration.[37] Russia has indicated several times that it would return two of the smaller islands to Japan in exchange for Japanese aid and other benefits. However, Japan insists that the four islands must be returned together. But Japan's bargaining power is very limited. Russia's need for Japan also varies from time to time, making the Russian position accordingly unpredictable.

Indeed, Russia's position has taken a hostile turn in the past year with two visits by Dmitry Medvedev in November 2011 as Russian President and in July 2012 as Prime Minister to those islands. Following Medvedev's visits, the Russian Defense Minister and his deputies also came to the islands to review Russia's defense posturing at this front line. By making these high-level official visits to the disputed islands, Russia has reaffirmed the legitimacy of its hold on the territory.

We do not know exactly why the Russians made these heavy-handed moves at this time. It appears that the change of geostrategic circumstances in Asia-Pacific, namely, the U.S. strategic shift toward this region, has influenced Russia's focus of attention. The connection is apparently China. Shortly before his visit to the disputed islands, the Russian President was in China to commemorate the 65th anniversary of the end of World War II. In the joint statement between Dmitry Medvedev and Hu Jintao, the two heads of state praised Russian and Chinese sacrifice in World War II, denounced unspecified attempts to alter the history of World War II and the key documents on post-war settlement, and vowed to preserve the hard-earned victory. The message is clearly directed at Japan and the United States.

Russia was happy to have China's support for its hold on the disputed territories with Japan. In joining hands with China, Russia has also expanded its reach into the Asia-Pacific region. China, for its part, invited Russia in with an eye on its dispute with Japan over the Senkaku/Diaoyu Islands in the East China Sea. China insists that the World War II settlement documents, such as the Cairo Declaration and the Yalta and Potsdam Proclamations, support China's claim to those islands. By getting Russia to reaffirm those World War II documents and accusing Japan of violating arrangements, China has strengthened its position against Japan.

While the two partners of convenience have found new common ground to advance their collaboration, their moves also cast a shadow on the U.S. strategic shift and military rebalancing toward Asia. These are developments that the United States definitely does not want to entertain.

The South Korean-Japanese Dispute.

This dispute is over a group of small islands in the middle of a sea between Japan and South Korea. The two nations have their own names for the islands and the sea: "Takeshima" in the "Sea of Japan" and "Dokdo" in the "East Sea" respectively. (See Figure 7; the English name for the islands is the Liancourt Rocks.)

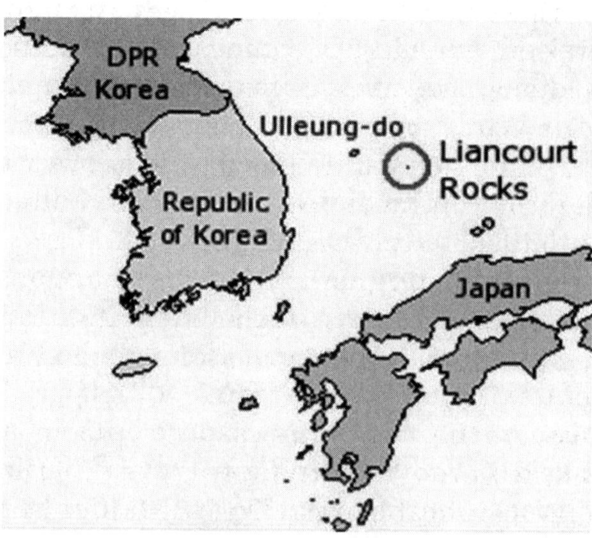

Figure 7. Japan-South Korea Territorial Dispute.

Japanese and South Koreans have written volumes of historical accounts about their claims to the islands. It will take pages here just to provide a glimpse of the complexity of their opposing views. However, for this analysis, it is critical to note that the most contested point in the dispute stems from the ambiguous treatment of Japan's occupied territories in the 1952 San Francisco Treaty of Peace with Japan.

The controversies are mostly in Chapter II, Articles 2 and 3 of the Treaty. Two key aspects are particularly relevant to the dispute between Japan and South Korea, and for many other territorial contentions in the Western Pacific, such as the ones between Japan and China and in the South China Sea. The first problem is that when Japan "renounces all right, title, and claim" to its once-occupied territories, it makes no reference as to whom those renounced territories are returned.

The other trouble is that the Treaty does not have a complete list of the islands in question (or not in question, for that matter). This "miss" was apparently not significant at the time. For one plausible reason, the participants at the peace conference did not have the means to make a complete account of the Japan-related islands in the Western Pacific. For another reason, they might not find the need to do that either—many of the "islands" are practically rocks barely seen when the tide is high, and they are too small and insignificant at a time when most of the international conflict of interest took place between and among nations on land but not in the ocean.

However, the two "misses" leave room for dispute. South Korea, for instance, argues that in addition to its historical claims, Dokdo should be part of those territories Japan renounced. Japan, on the other hand, insists that Takeshima is not mentioned in the 1952 Peace Treaty.

Whatever the case, South Korea has the upper hand over the dispute. It has had effective control since the early 1950s. Over the years, South Korea has also put many permanent structures on the islands. Moreover, South Korea also frames the dispute as part of the two nations' unsettled problem of history. Thus, the fight for these islands has regularly caused nationalistic and diplomatic frictions between the two nations. On August 10, 2012, South Korea's President Lee Myung-bak made a historic visit to Dokdo. By making a presidential visit to Dokdo, South Korea has significantly reinforced its effective control of the disputed islands. This act was also followed by a spat of unapologetic hostile exchanges with Japan. South Korea made it clear that there was no room for negotiation or compromise. But Japan is not going to give up its fight for those islands. This sticky issue between Japan and South Korea will continue to trouble the two allies for years to come.

The United States is torn between these two long-time allies and disappointed that its strong relations with these allies cannot get the two to bury their hatchet. However, the United States can do little to help settle the problem. It can only ask South Korea and Japan to "work out the dispute peacefully."[38]

The China-Japan Dispute.

This is undoubtedly the most explosive territorial dispute in Northeast Asia. It has been a contentious issue between China and Japan since the early 1970s. As presented later in this section, because the United States was involved in the making of the China-Japan dispute over the Senkaku/Diaoyu Islands and has commitment to the defense of Japan, this dispute has much broader consequence. As the crisis between China and Japan in 2012 over the islands in the East China Sea indicates (see details below), the U.S. rebalancing toward the Asia-Pacific may have already had its first test case in the making.

China-Japan Disputes over the EEZs and Continental Shelves.

China and Japan have two closely related disputes in the East China Sea. One is about the delimitation of the two nations' maritime boundary; the other, the sovereign right over the Senkaku/Diaoyu Islands in the disputed area of the two nations' overlapping ocean claims. It is very difficult, if not impossible, to see the settlement of one without the other.

China and Japan are maritime neighbors on the two sides of the East China Sea, with China's eastern seaboard from Fujian Province to Shanghai on the west and Japan's Ryukyu island chain on the east. The distance between the two sides is about 360 nm at its widest stretch in the north and about 200 nm at the narrowest points in the south. For centuries, there was no maritime boundary between China and Japan. However, with the birth of the UNCLOS, the two nations, which are parties to the treaty, found the need and requirement to establish proper dividing lines in their shared waters and the seabed underneath.

The UNCLOS offers two key provisions for the redistribution of the world's ocean commons. First, it encourages ocean littoral nations to claim 200-nm EEZs off their territorial waters. Second, it allows ocean littoral nations with naturally extended underwater continental shelves to expand the jurisdiction of their continental shelves to a maximum of 350 nm from their seashores.

These "revolutionary" provisions, however, were bound to create overlapping claims and bring neighboring littoral nations to confront each other. During the long and exhausted negotiations for the Law of the Sea, nations were divided on how to handle inevitable conflicts resulting from overlapping claims. Some advocated a one-size-fits-all "median line" to settle overlapping claims. Others insisted on an "equitable principle" for claimants to negotiate solutions to their disputes. The two sides could not reach an agreement at the conclusion of the UNCLOS in 1982. They compromised by referring claimants to follow the rules of the International Court of Justice to settle their disputes; and whatever method they use, they should have a formal agreement on the delimitations.

The UNCLOS came into effect in November 1994. Two years later, Japan promulgated its *Law on the Exclusive Economic Zone and the Continental Shelf*, which claimed a 200-nm EEZ all around Japan and asserted the use of median lines to delimitate overlapping claims with its ocean neighbors on the opposite sides.[39] China adopted its *Exclusive Economic Zone and Continental Shelf Act* in 1998 and claimed 200-nm EEZs along China's coast lines and its offshore islands as well.[40] China and Japan's claims in the East China Sea unsurprisingly overlapped. China was upset that Japan had taken the initiative to assert a median line as the delimitation in the two nations' expected overlapping claims. China's main objection was that it saw a natural prolongation of China's underwater continental shelf from its eastern seaboard stretching all the way to the Okinawa Trough, and therefore China was entitled to extend its jurisdiction over the continental shelf and use the western edge of the Okinawa Trough as a natural delimitation line between the two nations' claims.

(See the two nations conflicting delimitation lines in Figure 8; there is no precise measure of China's claim, but it is close to the 350-nm limit at its widest stretch.) From China's claim, there was no ground for Japan's asserted median line.

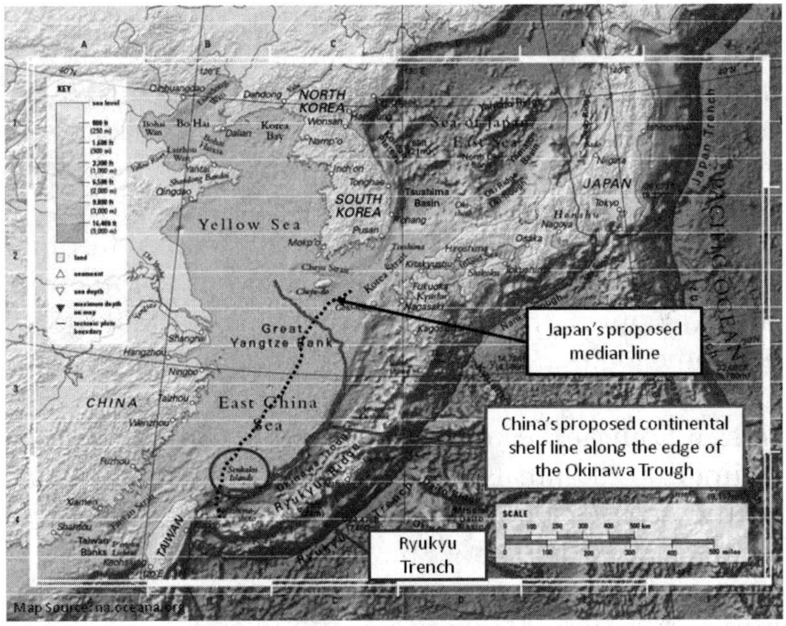

Figure 8. China-Japan Disputes in East China Sea.

Moreover, China took Japan's assertion as a unilateral act, deemed it invalid without an agreement between the two nations, and therefore dismissed it altogether. Although China and Japan subsequently held negotiations, their differences were oceans apart and no agreement was reached. This dispute has evidently affected the two nations in their efforts to explore natural resources in the disputed area (China's drill for natural gas and fossil oil near the alleged median line and Japan's protest is a case in point) and the overall China-Japan relations from time to time (when tension flairs up in the disputed area).[41]

The Fight over the Senkaku/Diaoyu Islands.

Complicating the dispute on the maritime delimitation was the two nations' fight over the sovereign ownership of the Senkaku/Diaoyu Islands in the disputed area. Those islands consist of three tiny uninhabited islands and five barren rocks that are barely visible on the ocean surface. They lie at the edge of the East China Sea continental shelf and the southern tip of the Okinawa Trough. (See the circled area in Figure 8.) The islands by themselves have little material value. However, they bear high-stake political and economic consequences for China and Japan.

Indeed, the current fight over these islands came initially out of the two nations' reaction to the speculation that the East China Sea had large fossil deposits,[42] and the expectation that the possession of those islands lends claim to a sizeable portion of the undersea natural resources. (According to one study, the area is about 20,000 square nm.[43]) In fact, Japan's median line delimitation was based on its assertion that those islands belonged to Japanese, and the islands were entitled to have an EEZ and continental shelf as well. If China were to "recover" the Senkaku/Diaoyu Islands and had its way on the extended continental shelf, China would have that valuable asset instead. (See Figure 8.) This is a fight neither side can afford to give up.

But the political reasons for the two nations' fight over those islands are equally significant. China and Japan are battling over those islands for their unsettled past, as well as for their unfolding future—

both want to be great maritime powers; and the fight over their maritime boundary and the possession of those islands is a test case of their ambitions. For China, moreover, it is also a fight with the United States over its alleged role in creating this dispute between China and Japan, its commitment to support Japan if China and Japan were to use force against each other over those islands, and a test of strength between China and the United States in their power transition process.

The stakes are high. The dispute is inexplicable. A list of the opposing arguments follows.[44]

- China holds that Chinese were the first to discover those islands and used them as navigation reference for centuries. Chinese fishermen came to the area around the islands regularly, and dynastic China's envoys made stops at those islands on their way to China's vessel state, the Ryukyu island kingdom, until Japan conquered the latter in 1879. China claims that those islands are "intrinsically integral and inseparable territories of China since antiquity" and "China's sovereign ownership of those islands indisputable."
- Japan does not dispute China's historical claims, but argues that China has never exercised effective control of those islands, and the "physical connection" of those islands to China is questionable, but not intrinsic. Japan, for the record, had official takeover and control of those islands from 1895 to 1945 and since 1972.
- China claims that the Senkaku/Diaoyu Islands are part of China's extended continental shelf and Taiwan's surrounding islands; Japan "stole" Taiwan and its surrounding islands,

including the Senkaku/Diaoyu Islands, from China through the Treaty of Shimonoseki of 1895; all of those islands therefore were covered in the Cairo Declaration and the Potsdam Proclamation and should be returned to China. (See Appendix 1 and 2.)

- Japan argues that it acquired the Senkaku/Diaoyu Islands through a cabinet decision prior to the Treaty of Shimonoseki as a *terra nullius*, land that is not claimed by any person or state; they were not mentioned in the Cairo Declaration and Potsdam Proclamation, and Japan argues it is therefore not required to "return" them to China. (See Appendix 1 and 2.)
- China condemns the United States for its inclusion of those islands in the trusteeship in 1951[45] and the handover of those islands to Japan in 1972. China charges that the U.S. policy was an ill-willed act toward China and a brute ill-treatment of China's territorial integrity.
- China holds that the Okinawa Trough marks the end of "China's underwater continental shelf," and the depth of the Okinawa Trough meets the UNCLOS requirement to be taken as a break in its extension.[46]
- Japan argues that the Okinawa Trough is only an accidental dent, and the Ryukyu island chain is the true edge of the East China Sea continental shelf. (See Figure 8.) Japan and China therefore share this continental shelf, and the two nations should delimitate their maritime boundary at the median line (with Japan continuing its possession of the Senkaku/Diaoyu Islands).

- China holds that it has followed a policy of "shelving disputes while promoting joint development" with good faith and accuses Japan of taking advantage of China's self restraint and altering the status of the disputes.[47]
- Japanese officials, most recently, Minister of Foreign Affairs Seiji Maehara, call the Chinese policy a "one-sided stand" and insist that "there is no territorial dispute" between Japan and China.[48]

The United States has always held that it takes no position in the territorial dispute and insisted that the trusteeship and handover have no bearing on the sovereignty of those islands. Moreover, the United States has also made it clear that the dispute should be resolved peacefully and if Japan were to be attacked as a result of this dispute, the United States would honor its mutual defense treaty obligation to come to Japan's defense. From the recent remarks by Secretary of State Hillary Rodham Clinton, this position cannot be more unequivocal and forceful:

> Well, first let me say clearly again the Senkaku fall within the scope of Article 5 of the 1960 U.S.-Japan Treaty of Mutual Cooperation and Security. This is part of the larger commitment that the United States has made to Japan's security. We consider the Japanese-U.S. alliance one of the most important alliance partnerships we have anywhere in the world and we are committed to our obligations to protect the Japanese people.[49]

Confrontations in 2012.

The China-Japan contention over the Senkaku/Diaoyu Islands came to a head in 2012. Both parties undertook measures that had fundamentally changed the way the two nations defined the issue and dealt with each other. The confrontations have also brought deep setbacks in the two nations' relations. The flares started in April 2012 when Japan's ultra-nationalist and outspoken Governor of Tokyo, Shintaro Ishihara,[50] launched a campaign to raise money for the purchase of the three Senkaku Islands (as marked in Figure 9) in the name of the Tokyo municipality (there are five other small islets in the Senkaku/Diaoyu island group, but they are not involved in the fight this time).[51]

Figure 9. The Senkaku/Diaoyu Islands.

Ishihara was angry at China's unbending claims on the Senkaku/Diaoyu Islands and China's increasingly frequent intrusions into the Japanese-controlled 12-nm territorial water zone around the islands. He was also troubled with the Japanese central government's "weak responses" to China. Moreover, Ishihara contended that the United States had not been straightforward with Japan, its most important ally in Asia, on Japan's territorial integrity. Ishihara was determined to reverse these trends and found the right place to make his statements; at the Washington-based think tank, the Heritage Foundation. As Ishihara's aid put it, the Tokyo Governor knew that "the foundation is a hard-liner against China and has the ability to spread information."[52] Of note is that Ishihara was about to quit his job as Governor to establish a new political party and intended to compete in Japan's next lower house election. It would be Ishihara's attempt to force change in Japan's parliament. His rally in Washington was to prepare the ground for his controversial moves in Japan.[53] Ishihara surely got what he wished for. His call for donations quickly generated a flow of cash in Tokyo. In less than 2 months, the amount reached ¥1 billion (Japanese yen, about $123 million).[54]

Ishihara also put the Japanese central government under enormous pressure — it subsequently was compelled to act on the island issue. Indeed, Prime Minister Yoshihiko Noda had to confront his Chinese counterpart, Premier Wen Jiabao, during their meeting in Beijing in May 2012 (which Ishihara intended to disturb). While the Chinese Premier called Ishihara's attempt to purchase the islands as an act against China's core interests, Noda insisted that the islands are Japan's territories, there is no dispute over those islands, and Ishihara's proposed purchase is Japan's internal business.[55] The Noda government, however, worried that Ishihara's agenda could drive the conflict between Japan and China out of control. It subsequently decided to preempt the purchase of the islands with a higher cash offer and put the islands under the central government's control.[56]

The Japanese central government's move was intended to calm the situation and prevent Ishihara from using the island issue to create further controversies, such as building a seaport there, turning the islands into a fisherman's refuge and possibly an anti-China outpost. But the Chinese government did not appreciate this "considerate act" of the Japanese government. China held that the purchase and subsequent nationalization of the islands by the Japanese central government was a collaborated attempt (with Ishihara) to change the status of those islands. Moreover, the Japanese central government could not have picked a worse time to conduct this business. The purchase and nationalization of the islands took place only a few days before September 18—the day Japan launched its invasion of China in 1937 (thereby starting the East Asia part of World War II) and a day the Chinese want to establish as a "National Humiliation Day." Japan's act was expectedly met with large-scale Chinese demonstrations across China.

Ishihara was presumably pleased to stir up the controversies. But his acts had backfired. China's reactions were too much for Japan to handle. The three most significant ones were:

1. The Chinese State Ocean Administration sent maritime administration and surveillance vessels to patrol the Senkaku/Diaoyu Islands. Chinese officials on board also demanded that Japan's coast guards stay away from China's "territorial waters."

2. On September 19, 2012, the Chinese Foreign Ministry announced that China had decided to submit its Partial Submission Concerning the Outer Limits of the Continental Shelf beyond 200 Nautical Miles in the East China Sea to the Commission on the Limits of the Continental Shelf under the UNCLOS.

3. The Chinese navy made warship navigation around the Senkaku/Diaoyu Islands.

China's first act had been going on for some time. The new development was that prior to this confrontation, the Japanese coast guard demanded the Chinese intruders stay away from the 12-nm zone around the islands; this time, the positions were reversed. The Chinese State Ocean Administration now had its maritime administration and surveillance ships make regular patrols inside the 12-nm zone around the Senkaku/Diaoyu Islands. They also demanded that the Japanese coast guard ships leave the "Chinese territorial waters."

Chinese have long argued that they have learned their lessons on the disputed territories in the Western Pacific the hard way and vowed to take "correct" measures to assert China's interests. In the words of Sun Shuxian, the Executive Deputy Commander of China's National Maritime Surveillance Fleet:

> [I]n international law, there are two customary practices for ruling on maritime disputes: one is to see if you have effective control and management of the disputed territory; and the other is the preference of effective control over historical claims. For instance, we have been arguing that these islands have been ours since antiquity; these words are hollow; what really counts is your actual control and effective management. China's marine surveillance and law enforcement patrol must make its presence in the disputed area and establish records of effective control.[57]

Sun's remarks were China's battle cries. They were followed by aggressive acts. The Chinese State Ocean Administration made the first attempt to break

Japan's control of the Senkaku/Diaoyu Islands in December 2008. Two Chinese marine surveillance and law enforcement vessels caught the Japanese coast guard off guard. They broke into the Japanese-guarded 12-nm zone around the Senkaku/Diaoyu Islands and stayed there for about 8 hours.[58] The Chinese Foreign Ministry spokesman dismissed Japan's protest, saying that:

> China does not see its normal surveillance and law enforcement activities in its maritime territory 'provocative;' and China will decide when to send these vessels to the Senkaku/Diaoyu Islands again on its own terms.[59]

Since then, China's maritime surveillance ships have made many more "visits" to the Senkaku/Diaoyu Islands. The Chinese State Ocean Administration has stated that it would turn these "visits" into regular official duties for the Marine Surveillance Fleet.[60]

In China's 12th 5-Year Plan (2010-16), the Chinese government has also planned to turn China's maritime patrol force into a formidable one. Funds have been earmarked to build 30 to 50 large and highly capable vessels. The first one, *Yuzheng* 310 (*Fishing Administration* 310), a 2,580-tonnage vessel with a platform for two Z-9A helicopters and advanced satellite communication systems, made its maiden voyage on November 16, 2010. Its destination was the Senkaku/Diaoyu Islands.[61]

China's advance in maritime patrol has drawn concerns in Japan. Many fear that if Japan does not take timely countermeasures, its maritime patrol forces will soon be no match to those of the Chinese. This is an alarming prospect for Japan.

China's second act was also devastating to Japan. If China's delimitation of its extended continental shelf stands, its control of the waters and seabed would expand to the edge of the Okinawa Trough. The Senkaku/Diaoyu Islands would be inside China's delimitation. This change would remove the basis for Japan's hold of the islands and its fight for the disputed oil fields discussed in the preceding pages.

China's third act was alarming. The Chinese navy has clearly indicated that it has the capability to support the Chinese Ocean Administration's efforts to gain control of the Senkaku/Diaoyu Islands. The seven Chinese navy warships were out on a training mission in the Pacific, sailed through the Okinawa Islands, and were only about 200 nm from the Senkaku/Diaoyu Islands.[62]

The United States has watched the China-Japan confrontations with great concern. Secretary of Defense Leon Panetta made a timely visit to Tokyo and Beijing amid the confrontations between China and Japan in September 2012. He "urged China and its neighbors not to engage in 'provocative behavior' over disputed islands and maritime claims, warning that it could escalate into a regional conflict that might draw in the U.S."[63]

As the confrontations continued to flare up, Deputy Secretary of State William J. Burns made the second round of U.S. senior official visits to Tokyo and Beijing in October 2012, an intense diplomatic effort in less than 1 month. But the U.S. officials had an earful of opposing views at the two East Asia capitals. Japan wanted the United States to support its sovereign ownership of the Senkaku Islands, only to find the United States reiterating its neutral position on the dispute, as in the words of Burns:

The United States, as you know, does not take a position regarding the question of the ultimate sovereignty of the Senkakus. But what we emphasize, very strongly, is the importance of taking a calm, measured approach to this issue; to focus squarely on dialogue and diplomacy; and to avoid coercion or intimidation or use of non-peaceful means.[64]

Japan was apparently upset with the ambiguous stand by the United States. Indeed, the United States not taking a position on the dispute is in essence a veto on Japan's claim. In this particular case, no position is a position. As the Japanese Ambassador to the United States complained, "The U.S. Government cannot be neutral over the Senkaku Islands." He insisted that the United States clarify its position.[65] China, on the other side, took the opportunity to challenge the United States to honor its "not-taking-position" policy and ask the United States to stay away from the disputes.

As it stands now, although the confrontation has not turned into a crisis, it has the potential to become one in the future. China has apparently gained an upper hand with its three moves mentioned above. At this moment, China's strategy is to press Japan to admit that there is dispute on the islands. In the long run, China will try to take the islands away from Japan. In fact, this long-term strategy is already underway. There is good reason to expect that a showdown between these two great Asian powers is only a matter of time.

Winners and Losers.

The conflict among the great powers in East Asia is taking a toll on regional welfare. The first victim is the 2012 APEC summit in Russia's Far Eastern city of

Vladivostok (September 7-8). Nothing significant was accomplished in this year's meeting. Secretary Clinton had to use the forum to address the disputes among the great powers. Prior to her arrival at the APEC, Clinton was hopeful that the Pacific would be big enough for the United States and China (the Secretary's remarks at the Pacific Islands Forum, Cook Islands, August 31, 2012). But facing Russia's "strategic shift to Asia" and the Russo-Chinese coalition, and frustrated with the great power disputes, the Secretary appeared to have second thoughts.

The second and more alarming consequence of this intensified great power struggle is the region's security and stability. While the dispute between South Korea and Japan is not likely to become armed clashes between the two allies, the ones involving Russian and China have significant military implications. Russia has recently reinforced its military deployment in its Far East front. China commissioned its first aircraft carrier in September 2012. In the meantime, China-Japan clashes around their disputed islands have already escalated from small fishing boat bumps to bigger-tonnage patrol vessel standoffs, e.g., Chinese State Ocean Administration vessels vs. Japanese coast guard ships. The Chinese navy has even sent warships to circle the troubled waters. Accordingly, the United States has taken conflict prevention measures by calling China to scale down the confrontations and by conducting aircraft carrier-centered military exercises in the Western Pacific to demonstrate the U.S. intervention capability.

These "titanic moves" are likely to cause the great powers to reconsider their security policies. There will be repercussions in the greater Asia-Pacific area as well. Vietnam, the Philippines, and Malaysia, the nations that have territorial disputes with China in the South China Sea, are taking measures to upgrade their defense capabilities. The Asia-Pacific region, which has been known for decades for its member nations' relentless pursuit of economic development and preference for peace, is now compelled to pay more attention to the security issues. To uphold peace and stability in this region, which are vital to U.S. interests, the United States should intensify its strategic rebalancing toward Asia-Pacific. While the United States should put all of its foreign policy instruments into full play, the U.S. military carries a special weight in this regard. The U.S. Army, along with the Pacific Command, should deepen its theater cooperation and engagement programs with all the actors in this region and military-to-military exchange with the Chinese military in particular. An effective engagement, supported by a strong U.S. military commitment, is the ultimate guarantee for peace and stability in this region.

U.S.-CHINA COMPETITION
IN SOUTHEAST ASIA

Southeast Asia is an area with great strategic significance. It sits at the crossroads of the Pacific and Indian oceans where some of the world's most important sea lanes of transportation are located. The Strait of Malacca is the lifeline of Japan, South Korea, China, and other Asian nations. The United States also uses the sea lanes extensively. With 10 nations and close to 600 million people, a pool of $1.5 trillion GDP, a landmass of over 4.6 million square kilometers, and a vast ocean stretch of over 7.5 million square kilometers, this region is what the geostrategy writer Nicholas J. Spykman describes as a core area of contention for great powers.[66] Indeed, for centuries, Southeast Asia has been a subject of outside control and influence. Chinese, Indians, Arabs, Europeans, and Americans have all left their footprints in this region. As a result, Southeast Asian nations have diverse political, cultural, and economic systems with different legacies.

In the post-Cold War era, Southeast Asian nations have made marked changes toward free market and democracy. The United States has a strong interest to encourage these changes and further integrate this vast and diverse region into the U.S.-led international order. Maintaining a strong relationship with Southeast Asian nations fits into the U.S. grand strategy and the Asia-Pacific strategy. Specifically, this relationship can help the United States in dealing with Islamic extremist terrorism and the emergence of a great power challenging U.S. supremacy; spreading democracy; promoting alliance of common values; controlling strategic chokepoints, resources, and markets; and developing a U.S.-led Southeast Asia regional order. The Obama administration is right on the mark to make this region a focal point in the U.S. "return to Asia." China, on the other side, has a very different policy calculation. It wants Southeast Asia to be a region friendly to China, a big market for China's economic development, and a stable place for China's security.

China has a long history of contact with and influence on Southeast Asia. (One simple data point puts it in perspective: about 50 million Chinese live overseas in the 10 nations of Southeast Asia, and the spilling effect is beyond words.) However, Chinese presence waned due largely to China's internal corruption, economic backwardness, and foreign invasions in contemporary times. China's security and economic prosperity along its southern borders had also deteriorated accordingly in the late-19th and early-20th centuries.

When the new China was founded in 1949, Chinese leaders attempted to restore contacts with the Southeast Asian nations and regain influence in the region. However, China's advance was minimal. Among the clearly decisive key factors impeding China's short-lived efforts were the worldwide Cold War between the United States and the Soviet Union; after all, China was on the Soviet side, whereas many of the Southeast Asian nations were U.S. allies. Mao's radical communist movements in the 1960s and 1970s at home and abroad, allegedly in some of the Southeast Asian nations (e.g., Indonesia), had further jeopardized China's relations with Southeast Asia and alienated China from this region again.

In 1978, China turned its attention to economic reform and modernization. This mission needed a war-free and stable environment for its development. But at the time, China had two big problems with its neighbors. One was a conflict-laden overall relationship with most of them. The other, which was also a contributing factor to the first, was the issue of many unsettled territorial disputes with its neighbors. Southeast Asia was a prime case in both of China's problems.

With near-collapse conditions at home, China at that time had no choice but to adopt a two-pronged approach to deal with the dire situations in its surrounding areas. The first leg of this approach was a "good neighbor" policy. The other part was a proposal to the disputing neighbors to shelve the territorial disputes for the time being, especially for those disputes that had no attainable solution for settlement in sight. With these two approaches as guiding principles, China was able to improve the conditions in its near abroad areas. This two-pronged policy is especially relevant to the Southeast Asia region. By the early 1990s, China had restored normal relations with all the Southeast Asia nations. In the last 2 decades, China has greatly expanded its economic relations in this region. In the 1997 Asian financial crisis, China stood firm to prevent the crisis from escalating into an uncontrollable catastrophy, hence establishing itself as a key player in the region's economy, especially Southeast Asia's economy.[67] In 2002, China was able to take the lead to create a ASEAN-China Free Trade Area (ACFTA). When the ACFTA came into effect in January 2010, it was the largest free trade area in terms of population and third largest in terms of GDP contribution in the world.[68]

In addition to the region-wide economic development, China has also engaged in some sub-regional projects such as the Greater Mekong River Regional Cooperation Operations, Balancing Economic Growth with Environmental Protection, and so on. While making headway in economic relations, China has gradually become a full dialogue member to almost all the Southeast Asian political, economic, and security regimes. As it stands now, China is practically a factor that no Southeast Asia agenda can proceed without taking into account.

China has long hoped that its approach toward the Southeast Asian nations should bring it benefits of: 1) a stable environment along its southern borders conducive to China's modernization mission; 2) a big marketplace for China's economic development; 3) a favorable condition to settle the territorial disputes in the South China Sea; 4) an improvement of China's image as a rising great power; and, 5) a testing ground for China to demonstrate its great power capacities and potentials. In all fairness, China has scored well in the first two expectations. However, China's approach has not brought expected returns for the last three items. Indeed, in spite of the previously mentioned improvements, China still finds the "China threat" outcry echoing in Southeast Asia from time to time. Although China has kept reassurring the Southeast Asian nations that China is no threat to them, it continues to see the Southeast Asian nations courting the United States and seeking comfort from U.S. security arrangements. The Chinese were definitely upset when they saw the Singapore revered statesman Lee Kuan Yew's public appeal for the United States to maitain its presence in Southeast Asia and counterbalance China's influence in the region.[69] Even worse, China finds it especially disturbing that its decades-long efforts have not led to the reduction or settlement of the territorial disputes in the South China Sea; instead, the conflicts have intensified.

China has much to consider in reference to its current relations with the Southeast Asian nations. Yet there are several key factors with which Chinese leaders cannot squarely come to terms. The first problem is that they refuse to admit that an ideological divide is standing in the way of China's attempt to promote a truly friendly relationship with the Southeast Asian nations. As the Southeast Asian nations move further down the road of democracy, they find less and less common language with China in governance, universal human rights, and the true rule of law. Many may argue that these are smokescreens. But one cannot deny that cooperation based on common interests alone is only a matter of convenience. At the end of the day, common values hold true friends together.

The second problem is that China tends to forget that it is too big for the Southeast Asian nations, and its national power is too much for any nation in the region to deal with individually. In such a situation of disparity, it is natural for the Southeast Asian nations to rely on ASEAN or to turn to an outside great power like the United States for counterbalance against China. It is a strategy any small nation will choose. It is notable that while courting the United States, the Southeast Asian nations also try to avoid becoming an overt U.S. instrument to counterbalance China. In other words, Southeast Asian nations do not passively react to the great-power competition; they try to get the best possible benefits out of the competition between China and the United States. They have long pursued and played well a dual strategy called by a Southeast Asia specialist "omni-enmeshment" and "balance of great power influence."[70]

Finally, the Chinese take it for granted that their claims on the South China Sea islands are indisputable and all the other disputants "have stolen" islands from China. The Chinese believe that they will eventually have the power to reclaim those "stolen" territories, and the other disputants should have no illusion about that. Given this presumably no-win outcome, the other disputants might as well make the best out of their "holding" while they can. The Chinese therefore expect that the economic benefits (or bribery in China's corruption cultural sense) should get the other disputants to soften their stands on the disputed territories in the South China Sea.

SOUTH CHINA SEA DISPUTES

The South China Sea encompasses a portion of the South Pacific spanning from the southern tip of Taiwan to the Strait of Malacca. The area includes numerous small islands, rocks, and reefs scattered roughly around the four island groups known as the Pratas in the northeast, the Macclesfield Bank in the middle, the Paracel Islands in the west, and the Spratly Islands in the south (see Figure 10). Many of the "features," however, are submerged under water, visible only during low tides. There is, therefore, no precise count of the features in the South China Sea.

China has a long history of fishing in the area surrounding these islands, and their official reach goes back nearly as far. The Chinese were arguably the first to assign them names, use them as navigational references, and attempt to designate them as Chinese territories by putting them in the jurisdiction of southern Chinese coastal provinces and marking them as such in maps.[71] For centuries, the Chinese took it for granted that their historical reach established their ownership over those islands and the waters around them. They never felt the need to maintain effective control or management of those faraway and uninhabitable islands. This was not a problem when the Middle Kingdom was powerful and its influence on its surrounding areas was strong.

Yet when China was on dynastic decline, which has been a "cyclical illness" of China throughout its history, its imperial reach also retracted. China's latest dynastic decline met with the forceful arrival of the European colonial powers. This time, in addition to suffering from internal turmoil, China also "lost" practically all of its offshore "territories" (in quotation marks because they are in dispute) to the foreign powers: Taiwan and its surrounding islands were ceded to Japan; the South China Sea islands all "acquired" European names (the British were arguably the first Europeans to set foot on the South China Sea islands; indeed, the Spratly and Pratas Islands were both renamed after British sailors);[72] the French took possession of the Paracel and Spratly Islands in the 1930s to expand the reach of its colonial protectorate Annam (the predecessor and central region of present day Vietnam); and during World War II, Japan took control over all of the South China Sea islands in its drive to create the Greater East Asia Co-Prosperity Sphere.

At the end of World War II, Japan complied with the demands by the U.S.-led allies, as articulated in the Cairo Declaration (1943) and reaffirmed in the Potsdam Proclamation (1945) to relinquish all the territories it "had stolen" during its imperial expansion (see Appendices 1 and 2). However, by the time Japan came to sign a peace treaty with its wartime opponents and victims to legalize the termination of war and its relinquishments, there was no undisputed recipient to accept the territorial "spoils." China was divided between two governments, each claiming to represent the whole. The national leaders gathering in San Francisco for the peace conference with Japan could not decide which China, the ROC on Taiwan or the PRC on the mainland, should be designated as the legitimate recipient of Taiwan and its surrounding territory. In fact, neither Beijing nor Taipei was invited to the conference. In the end, the Peace Treaty with Japan only reiterated Japan's renunciation of its right to Taiwan and Pescadores but did not specify the recipient. With respect to the South China Sea islands, the delegates to the peace conference rejected a Soviet proposal to give them to China[73] and did not endorse a claim by Vietnam at the conference.[74]

China denounced the design of the peace treaty with Japan as well as the outcome of the San Francisco peace conference.[75] Chinese Foreign Minister

Zhou En-lai issued a statement prior to the conference condemning the United States for its alleged role in "depriving China of its right to recover its lost territories" and "creating a treaty for war but not peace in the Western Pacific." At the same time, China reiterated its claim to Taiwan, its surrounding islands, and all of the South China Sea islands.[76]

In retrospect, China had several opportunities to secure its claim and control of the South China Sea islands regardless of what the United States and other nations did at the peace conference in San Francisco. In 1943, and in a world still heavily ruled by "jungle power" (in the way of the centuries-old power politics, great powers did what they wanted, but small nations suffered what they must, and great powers got to decide post-war international order), China could have demanded the "return" of the South China Sea islands in the Cairo Declaration. Indeed, the United States, the United Kingdom, and China were the only three "Great Allied Powers" gathering in Cairo to map out the post-war East Asia territorial rearrangement. (Vietnam, Malaysia, and the Philippines were not even independent countries yet.)

Moreover, in 1946, the ROC government dispatched warships to "recover" the Paracel and Spratly Islands.[77] In a world that emphasized effective control rather than historical claims,[78] China could have kept its troops there to exercise effective control of those territories and establish China's unbroken and unchallengeable possession of those islands. Chinese leaders are themselves to blame for failing to do so and neglecting the South China Sea islands decades thereafter.[79] Their repeated protests against the United States and the other claimants and their statements about the South China Sea islands "historically belonging to

China," or as "China's intrinsic and inseparable territories," although necessary for China to uphold its claims, sounded painfully hollow.[80] Chinese leaders wasted all their time and energy engaging the Chinese in "perpetual revolution and class struggle" against each other at home while leaving the disputed territories unattended offshore.

In the meantime, Vietnam and the Philippines continued their efforts (in acts, not only in words) to secure their claims and exercise effective control over the South China Sea islands.[81] By the early 1970s, word came that the South and East China Seas had vast deposit of fossil fuel and natural gas. The negotiation of the UNCLOS was also making progress—the world would soon divide up the "ocean commons" and allow the ocean littoral nations to claim the 200-nm EEZ's and take possession of their naturally extended underwater continental shelves. These new developments prompted the South China Sea littoral nations to "scramble for effective occupation" of the islands in the South China Sea.[82] This scramble for territory continued well into the 1990s and left the disputes on the South China Sea islands as follows:

- The Pratas Islands: completely occupied by Taiwan, but disputed by China;
- The Paracel Islands: mostly occupied by China, but disputed by Vietnam;
- The Macclesfield Bank: disputed among China, Taiwan, and the Philippines;
- The Scarborough Shoal: disputed among China, Taiwan, and the Philippines;
- The Spratly Islands: disputed among China, Vietnam, Taiwan, the Philippines, Malaysia, and Brunei; of the more than 30,000 features,

about 50 are considered islands; they are occupied by the following disputants:
— China: 6;
— Vietnam: 29;
— Malaysia: 5;
— Philippines: 9;
— Taiwan: 1;
— Brunei, none, but has EEZ dispute.[83]

In the face of these disputed claims, China continues to hold that it is the owner of all the islands, reefs, and other features in the South China Sea and accuses all others of "stealing and occupying China's territories." Vietnam holds the second largest claim. In addition to disputing China over the Paracel Islands, Vietnam claims ownership to all of the Spratly Islands. Its claim puts Vietnam in dispute with China, Taiwan, and its Southeast Asian neighbors Malaysia, Brunei, and the Philippines.

China was upset with the other claimants' rush to take possession of the South China Sea islands. It used force against Vietnam in 1974 to "regain" control of the key parts of the Paracel island group and used force against Vietnam again in 1988 to fight for the islands in the Spratly group. There have also been armed conflicts between China and the Philippines over their disputed features.

While dealing with its neighboring disputants, China has also tried to prevent the involvement of the United States in these disputes. Throughout the years, China has been very suspicious and sensitive to the U.S. position on the South China Sea disputes.[84] China blamed the United States for making the sovereignty of the South China Sea open for dispute at the San Francisco peace conference in 1951. They were also upset with the United States for freely using the South China Sea to wage the Vietnam War (transporting forces and launching air and naval attacks on Vietnam), ignoring China's claim and protests, and making Southeast Asia and the South China Sea one of the three "anti-communism breakwaters" in the Western Pacific during the early years of the Cold War (through the Southeast Asia Treaty Organization [SEATO]; the other two are the U.S.-Japan and Korea alliances and the U.S.-ROC [Taiwan] defense pact).[85]

The Chinese were "grateful," however, when the United States took a hands-off stand on the South China Sea disputes following its rapprochement with China in 1972. For instance, the United States turned a blind eye to China's military operation against Vietnam in 1974 (China-Vietnam naval clash over the Paracel), 1979 (China-Vietnam Border War), and 1988 (China-Vietnam naval clash over the Spratly). But they got upset again when the United States took issue with China's military clash with the Philippines in 1994, warned of China's "creeping encroachment" of the South China Sea territory,[86] and hinted that the U.S.-Filipino defense treaty would cover the Philippines' claimed South China Sea territories.[87]

Chinese leaders have taken watchful notes of the U.S. adjustments in its position toward the South China Sea disputes since the end of the Cold War. Although the United States has openly maintained a neutral position,[88] China nevertheless holds that the United States privately sides with the Southeast Asian claimants.

More recently, the Chinese see growing U.S. domestic pressure on the U.S. Government to take stronger stands against China on the South China Sea disputes. U.S. anti-China critics strongly urge the

Obama administration to be more assertive in Southeast Asian affairs. They also charge that China's claim on the South China Sea islands is overbearing. They are concerned that China's military modernization is upsetting the strategic balance in Southeast Asia and threatening U.S. navigational freedom (such as the harassment of U.S. surveillance ships and flights). They press the U.S. Government to modify its strategy toward China in Southeast Asia and the South China Sea and urge the U.S. Government to support Vietnam, the Philippines, and Malaysia on their claims.[89] "The United States should take sides," as some in the United States demand.[90]

The Chinese note that the Obama administration appears to take those domestic pressures seriously. In less than 2 years since taking office, Secretary of State Clinton visited this region six times. She repeatedly told the Asia community that the United States is back (from the George W. Bush's "neglect") and is here to stay.[91] Secretary of Defense Robert Gates echoed Clinton's call by emphasizing the United States as a "residence power" in Asia and reaffirms U.S. commitments to this region.[92] President Obama also visited Asia twice and characterized himself as the "first Pacific President."[93] Through these high-sound-bite outreaches, the Obama administration put forward a strategy toward Asia: strengthen and reinvigorate old alliances, make new friends, and support multilateral institutions in this region.[94]

The Chinese watch the Obama team's moves with much suspicion. They dismiss the above as pretext for the United States to reposition itself in the Western Pacific. They argue that the United States has never left Asia-Pacific, even though it has been busy fighting wars elsewhere, and this stormy repositioning is only an attempt to counterbalance China's expanding power.[95] Thus, instead of welcoming the "return" of the United States to Asia, China was preparing for new tension in the two nations' relations.[96]

Unfortunately, it did not take China long to see a downturn this way. Indeed, an eventful 2010 unfolded in a series of confrontations between the United States and China that practically touched upon almost all the sensitive issues between the two nations. The most explosive ones were 1) the U.S. decision to sell $6.4 billion worth of arms to Taiwan and subsequent Chinese suspension of military-to-military exchange with the United States, and, 2) the U.S.-China test of will over the alleged North Korean sinking of a South Korean warship and its aftermath. (The United States wanted to send the *George Washington* aircraft carrier strike group into the Yellow Sea for a joint military exercise with South Korea against North Korea; China vehemently opposed the U.S. plan and eventually forced the United States and South Korea to conduct the exercise in the Sea of Japan.)

In March 2010, U.S. Deputy Secretary of State James Steinberg and the National Security Council's Senior Director for Asian Affairs Jeffrey Bader visited Beijing in an attempt to "bring U.S.-China relations back on track." Their meeting with the Chinese officials, however, was an unsuccessful one. Chinese officials took the occasion to lecture their American guests on China's core interests. But since the two sides did not see eye to eye on those issues, they could not agree on the way to handle the issues, and their differences remained as wide as ever.

It was later revealed that during that meeting, Chinese officials, for the first time, included the South China Sea territorial dispute in the list of Chinese core interests.[97] This Chinese move, even if it was meant to simply test the waters was very disturbing to the United States, for China has long held that it will use all instruments of national power, especially the use of force, to deal with issues involving its core interests. In addition, raising the stake on the South China Sea dispute is very dangerous. China has a broad claim on the South China Sea, not just land features, but also waters. Given China's position on foreign military activities in its claimed zones, putting the South China Sea as China's core interest has far-reaching consequences. Thus, 2 months later in May, when Chinese officials brought this issue directly to Clinton while she was in Beijing for the Second U.S.-China Strategic and Economic Dialogue, Secretary Clinton rejected it flatly: "We don't agree with that."[98]

The push and shove between the two nations came to a head in July 2010. The scene was in Hanoi, Vietnam, and the occasion was the annual ASEAN Regional Forum. Secretary Clinton came prepared to give China an official response on the South China Sea issues. She declared the following:

- The United States has a national interest in the freedom of navigation, open access to Asia's maritime commons, and respect for international law in the South China Sea.
- The United States supports a collaborative diplomatic process by all claimants for resolving the various territorial disputes without coercion. We oppose the use or threat of force by any claimant.
- While the United States does not take sides on the competing territorial disputes over land features in the South China Sea, we believe claimants should pursue their territorial claims and accompanying rights to maritime space in accordance with the UN convention on the Law of the Sea. Consis-

tent with customary international law, legitimate claims to maritime space in the South China Sea should be derived solely from legitimate claims to land features.[99]

The Chinese charged that, taken out of context, the above sounded righteous; but delivered at the ASEAN Regional Forum, every foreign minister in the audience (there were 27 of them at the forum)[100] knew what Clinton was after, and every point made in her speech was an attack on China.

Clinton's first point is a forceful statement. If it stands, this statement can become a doctrine in U.S. foreign policy on par with other U.S. foreign policy doctrines, most notably the Monroe Doctrine, that put the European powers on notice and defined U.S. interest in the Western Hemisphere, and the Carter Doctrine that warned the Soviets not to tamper with the Persian Gulf and put at risk the security of the region, a vital interest of the United States.[101] This "Hillary Clinton Doctrine" is put forward against another great power, China, and defines U.S. position on the key issues at stake.

Clinton's second point goes against China's long-held position of settling disputes with the other claimants in bilateral ways. The United States is concerned that China may have too many advantages over the the other disputants when considered individually. In addition, by opposing the threat or use of force in settling the South China Sea disputes, Clinton was in essence telling the Chinese that they should not make the South China Sea disputes a core interest of China.

Clinton's third point goes against another Chinese long-held position of settling the disputes "in accordance with special historical, political, economical, geographical, and other related circumstances." To the Chinese, the UNCLOS is a necessary reference, but they do not want to subjugate the disputes to the ruling of the UNCLOS, for it will be disadvantageous to China's claims, which are largely historical but not records of effective control.

Clinton's final point takes issue with an ambiguous Chinese claim on the South China Sea. It is the area delimited by the nine dashed border lines. China has had these dashed lines around the South China Sea on its maps since 1947, when the first map was published by the Chiang Kai-shek's Nationalist government shortly before its fall and retreat to Taiwan. However, neither the Nationalist government nor the PRC government has ever clarified whether those dashed lines are temporary markers of China's territorial boundary that cover both the water as well as the land features in the South China Sea and would be eventually formalized as permanent Chinese border lines. By taking an official stand on this issue, Clinton is dismissing those Chinese markers. *The United States is now a disputant in the South China Sea disputes.*

The Chinese were furious. They had asked Clinton not to bring this issue to the ASEAN Regional Forum prior to the meeting. They were angry that the U.S. Secretary of State ignored the Chinese request and took such a forceful stand at the forum. Chinese Foreign Minister Yang Jiechi immediately responded with "a very strong and emotional statement, essentially suggesting that this was a pre-planned mobilization on this issue. . . . He was distinctly not happy."[102]

China rejected the U.S. attempt to "internationalize and complicate" the South China Sea disputes and vowed not to cave in to the U.S. pressure. In an unmistakable show of its resolve, China had the PLA carry out a large-scale live-fire military exercise in the South China Sea, reportedly involving all of China's naval fleets (the Northern, Eastern, and South China Sea fleets), shortly after this confrontational exchange.[103]

The PLA naval exercise was also an apparent countermeasure against the upcoming first-ever U.S.-Vietnam military exercises in the South China Sea. The U.S.-Vietnam military exercises were to commemorate the 15th anniversary of the U.S.-Vietnam rapprochement. But put in the context of this recent tension between China and the United States, China clearly interpreted it as part of the Obama strategy to make new friends and a U.S. effort to form a U.S.-Vietnam "united front" against China. The timely arrival of the *George Washington* carrier strike group (immediately following its joint exercises with South Korea in Northeast Asia) gave the Chinese solid evidence to support their views.

The Chinese see that the United States is abandoning its half-hearted neutral stand and moving toward an active involvement approach.[104] To the Chinese, this is like a nightmare come true—the last thing they want to have is a confrontation with the United States over the South China Sea disputes. Unfortunately, they see it becoming a reality. By any account, these open and subtle exchanges constitute a defining moment in the U.S.-China power transition. South China Sea disputes have also become a complicated part of this contentious process between China and the United States. The Chinese believe that this development is inevitable and beyond China's control.[105]

CONCLUDING REMARKS

This strategic assessment has highlighted the key characteristics of Asia-Pacific with respect to its economic development, political institution, and security situations. It should be painfully clear that the trends are indeed mixed and complex. Indeed, while there is great potential for Asia-Pacific to become the true powerhouse in the unfolding Pacific Century, there are also risks of conflict in many respects, the most destructive of which are arguably territorial disputes. On top of the mixed currents, one can clearly see the impact of the U.S.-China competition. For better or worse, the U.S.-China power transition is complicating the Asia-Pacific regional relations.

President Obama won his second term in office. Shortly after the general election, the President made a 3-day trip to "three strategically important Southeast Asian countries: old U.S. ally Thailand, new friend Myanmar (Burma), and China ally Cambodia, in a visit that underlines Washington's expanding military and economic interests in Asia under last year's so-called 'pivot' from conflicts in the Middle East and Afghanistan."[106]

More precisely, President Obama was there to attend the East Asia Summit. Although the visit was overshadowed by the armed conflict between Israel and Palestine, the President's message was clear: the United States will continue its strategic shift toward Asia-Pacific. The White House briefing on the outcomes of the summit puts it best:

> President Obama attended the East Asia Summit (EAS) on November 20 in Phnom Penh, Cambodia, as part of the Administration's continued focus on rebal-

ancing its engagement in Asia to reflect the economic and strategic importance of this dynamic region. As an Asia-Pacific power the United States' economic and security future is inextricably linked to the region, and President Obama used the summit to explore with other Asia-Pacific leaders ways to enhance cooperation on the region's most pressing challenges, including energy, maritime security, non-proliferation, and humanitarian assistance and disaster response. The President made clear that full and active U.S. engagement in the region's multilateral architecture helps to reinforce the system of rules, norms, and responsibilities, including respect for universal human rights and fundamental freedoms, that are essential to regional peace, stability, and prosperity.[107]

The stage is set. All nations in the Asia-Pacific will take the U.S. initiatives into account and follow their national interest to find exit strategies for those difficult dilemmas.

ENDNOTES

1. See an in-depth account of China's tortuous road to modernity in David Lai, *The United States and China in Power Transition*, Carlisle, PA: Strategic Studies Institute, U.S. Army War College, 2011.

2. One should also take China's gross domestic product (GDP) with a grain of salt. As Chinese Vice Premier Li Keqiang (李克强) puts it, China's GDP is "forged and not reliable" ("joking remarks" to U.S. Ambassador Clark T. Randt, Jr., in 2007, reported by the Chinese newspaper, *The World Journal* [世界日报], October 23, 2012). However, there is reasonable ground that as China becomes more integrated with the international system, its vital statistics should be more in line with the international standards. That said, the current measures have been used by the Chinese and the international community for years.

3. See World Bank, *World Economic Outlook Database*, International Monetary Fund Report, October 2012; and *CIA World Factbook*, Washington, DC: Central Intelligence Agency, June 2012.

4. A goal repeatedly articulated in the Chinese Communist Party's party platforms and China's 5-Year Plans as well.

5. The most critical measure is internal. When the leaders of a great power do not understand the logic of national power as a function of growing wealth and keep their nation in a condemned situation, there is little the outside world can do to help them. In fact, from the perspective of great power politics, it is better off for the rest of the great powers to "let this dog sleep." But when a great power's leaders come to their senses and set their nation on the path to economic development, there is little the outside world can do to stop it. Indeed, China can only be set back or defeated by the Chinese themselves. The key issue is whether the Chinese Communist Party (CCP) can keep itself in power and lead China to achieve the goal. This is a monumental mission. The CCP has managed to muddle through so far. There is a good chance that it can continue to do so in the years ahead.

6. For the seminal introduction of the power transition theory, see A. F. K. Organski, *World Politics*. New York: Alfred A. Knopf, 1958. For the most recent analysis of power transition, see Lai, *The United States and China in Power Transition*.

7. In recent years, China has been advocating a Confucian view of "harmonious world order" and setting up numerous Confucius colleges in the world to promote Confucian values. In 2004, Chinese President Hu Jintao put forward a new mission for the Chinese military, the People's Liberation Army (PLA), in the new century. The most significant aspects of this new mission is to turn on the green lights for the PLA to go beyond China's territorial boundaries to protect China's growing and expanding interests and to safeguard international peace and stability. Sound familiar? These are "patented" U.S. calls. They are now coming from China.

8. One should be careful not to confuse the U.S.-China power transition with interstate rivalries. The latter usually take place between two nations for limited scopes. Good cases in point are

the decades-long India-Pakistan rivalry, the Iran-Iraq rivalry in the 1980s, the rivalry between the two Koreas, and so on. These rivalries are not about the international system and order, as is the power transition between the United States and China.

9. There have been numerous writings about this subject matter in the last 2 decades. The alarming ones have no doubt reached the decisionmakers in Beijing and Washington.

10. The most comprehensive articulation of China's peaceful development comes from the mastermind of this idea, Mr. Zheng Bijian. He was a long-time advisor to Chinese President Hu Jintao and Vice President of the Chinese Communist Party Central Party Cadre School at the time of the conception of this call. See Zheng's two key publications: "China's New Path of Development and Peaceful Rise," speech at the 30th Annual World Forum, Lake Como, Italy, *Xuexi Shibao*, Beijing, November 22, 2004; and "China's 'Peaceful Rise' to Great-Power Status," *Foreign Affairs*, September/October 2005.

11. Robert B. Zoellick, "Wither China: From Membership to Responsibility?" Remarks to the National Committee on U.S.-China Relations, September 21, 2005, New York City. The 2006 U.S. National Security Strategy endorsed Zoellick's designation of China as a "responsible stakeholder" and reiterated the key points made by Zoellick.

12. Organski is the first to point out that when the leaders of a backward great power decide to modernize their nation, there is little the outside world could do to stop it, unless, of course, the other great powers wage a war to that end. See Organski.

13. For a discussion of the deadly contest between great powers as a result of power transition, see Thucydides, *The History of the Peloponnesian War*, New York: Penguin Books, 1954. For a discussion of the so-called "Thucydides Trap," see John J. Mearsheimer, *The Tragedy of Great Power Politics*, New York: Norton, 2001. See also Mearsheimer's other writings about the tragedy: "The Future of the American Pacifier," *Foreign Affairs*, September/October 2001; "Clash of the Titans: (Debate between Zbigniew Brzezinski and John J. Mearsheimer)," *Foreign Policy*, January/February 2005; and "China's Unpeaceful Rise," *Current History*, April 2006.

14. Steven Erlanger, "Christopher to Meet His Chinese Counterpart." *The New York Times*, March 20, 1996.

15. Comprehensive analyses of China's anti-access and area denial (A2/AD) capabilities can be found in the writings at the Center for Strategic and Budgetary Assessment. See Andrew F. Krepinevich, *Why Air-Sea Battle?* 2010; Jan van Tol with Mark Gunzinger, Andrew Krepinevich, and Jim Thomas, *Air-Sea Battle: A Point-of-Departure Operational Concept*, 2010. The RAND Corporation analysts have also provided a piece: Roger Cliff, Mark Burles, Michael Chase, Derek Eaton, and Kevin L. Pollpeter, *Entering the Dragon's Lair: Chinese Anti-access Strategies and Their Implications for the United States*, Santa Monica, CA: RAND, 2007. The Pentagon's annual report to Congress on China's military power has also documented China's development of A2/AD capabilities in the years since 2000.

16. U.S. Department of State, "Clinton on America's Pacific Century: A Time of Partnership," November 10, 2011.

17. Joe McDonald and Youkyung Lee, "China Overtaking U.S. as Global Trader," *Associated Press*, December 3, 2012.

18. For a recent development of the U.S. push for the Trans-Pacific Partnership (TPP), see Howard Schneider, "U.S., Asian Nations in Trade Talks without China," *The Washington Post*, September 21, 2012.

19. U.S. Department of Defense, "Sustaining U.S. Global Leadership: Priorities for 21st Century Defense," January 2012. U.S. officials maintain that the Air-Sea Battle concept is not designed against any designated adversary. However, it is painfully clear that there are only two targets. The primary one is China. The secondary and less formidable one is Iran. Thus far, only the Center for Strategic and Budgetary Assessment, the principle advocate for the Air-Sea Battle concept, openly and unmistakably acknowledges and addresses the China problem. But its calls are dismissed as unofficial and not representing the U.S. Government position. For the details of the Air-Sea Battle concept, see the Department of Defense, *Joint Operational Access Concept (JOAC)*, January 2012, and the Center for Strategic and Budgetary Assessment, *Air-Sea Battle: A Point-of-Departure Operational Concept*, 2010.

20. U.S. Department of State, "Clinton on America's Pacific Century: A Time of Partnership," November 10, 2011.

21. Hillary Clinton, "America's Pacific Century," *Foreign Policy*, November 2011.

22. International practice gives more weight to effective control than historical claims. See the Island of Palmas Case ruled by the Permanent Court of Arbitration in 1925 (a dispute between the United States on behalf of its colony the Philippines and the Netherlands on behalf of its colony Indonesia), which sets the example of effective exercise of sovereignty taking precedence over the right of discovery and claim of territorial contiguity (see reference Hague Court Reports 2d 83 [1932] [Perm. Ct. 4rb. 1928]). Other cases, such as the Clipperton Island, Eastern Greenland, and the Minqulers and Ecrehos Islands, all point to the same requirement of effective control over historical claims.

23. China Anti-Secession Law of 2005, available from *www.china.org.cn/english/2005lh/122724.htm*. For better or worse, those conditions are subject to China's interpretation.

24. John King, "Blunt Bush Message for Taiwan," CNN Washington Bureau, December 9, 2003. See also Thomas J. Christensen, Deputy Assistant Secretary for East Asian and Pacific Affairs, "A Strong and Moderate Taiwan," Speech at the U.S.-Taiwan business Council, September 11, 2007.

25. "Strategic ambiguity" on the Taiwan issue was an unwritten or spoken but loosely followed U.S. policy until President George W. Bush abolished it in 2003.

26. There have been many angry calls for the Chinese government to take measures to stop the U.S. sales of arms to Taiwan in the aftermath of the last Obama administration authorization in January 2010. See David Lai, "Arms Sales to Taiwan: Enjoy the Business While It Lasts," *Of Interest*, Strategic Studies Institute, May 3, 2010.

27. For an in-depth discussion of why China cannot set Taiwan free and the origins and evolution of the Taiwan issue, see Lai, *The United States and China in Power Transition*.

28. See Shirley A. Kan *et al.*, "China-U.S. Aircraft Collision Incident of April 2001: Assessments and Policy Implications," CRS Report for Congress, Washington, DC: Congressional Research Service Report, October 10, 2001; and Eric Donnelly, "The United States-China EP-3 Incident Legality and Realpolitik," *Journal of Conflict & Security Law*, Vol. 9, No. 1, 2004, pp. 25-42, for a discussion of the incident and disputes on military operations in the Exclusive Economic Zone (EEZ).

29. *Xinhua News Agency*, "U.S. Military Surveillance Ships to Collect Military Intelligence in Chinese Waters," *www.xinhuanet.com*. For these monitoring and patrolling acts, see also China Marine Administration Law Enforcement Annual Report 2006, available from *www.soa.gov.cn*; Jim Garamone, "Chinese Vessels Shadow, Harass Unarmed U.S. Surveillance Ship," March 9, 2009, available from *www.navy.mil*.

30. Admiral Blair made this remark at the Senate Armed Service Committee Hearing, reported by the Staff Writers, Washington (AFP), March 11, 2009.

31. Liu Juntao, "[Foreign Minister] Yang Jiechi Met with [Secretary of State] Hillary Clinton and Stressed [Chinese] Opposition to Foreign Military Force to Have Military Exercise in the Yellow Sea," July 23, 2010, available from *www.people.com.cn*; "Chinese Ask U.S. Military to Stop Surveillance in China's Exclusive Economic Zone," *People Daily*, August 28, 2009.

32. Huang Xueping , Chinese Defense Ministry Spokesperson, "Defense Ministry Spokesperson Press Remarks," Xinhua News Agency, March 11, 2009; Ma Chaoxu, "Chinese Foreign Ministry Spokesperson Press Remarks," Xinhua News Agency, May 6, 2009.

33. The UNCLOS allows member nations to hold reservations. See Under Secretary of Defense for Policy, "Maritime Claims Reference Manual," DoD 2005.1-M. June 23, 2005, for a documentation of the nations signing, ratifying, or rejecting the treaty. The 14 nations that hold reservations are Bangladesh, Brazil, China, Finland, Egypt, Guyana, India, Iran, Kenya, Malaysia, Nicaragua, Pakistan, Portugal, and Uruguay. See also Raul Pedrozo, USN Captain, JAGC, "Close Encounters at Sea," *Naval War College Review*, Vol. 62, No. 3, Summer 2009.

34. Author's conversation with senior U.S. military leaders. See Eric A. McVadon, "The Reckless and the Resolute: Confrontation in the South China Sea," *China Security*, Vol. 5, No. 2, Spring 2009, for an excellent analysis of this issue.

35. U.S. Pacific Command, "Press Roundtable in Beijing, China by Admiral Timothy J. Keating," Commander's Transcript from January 17, 2008, available from *hongkong.usconsulate.gov/ustw_others_2008011701.html*.

36. See Pentagon's Annual Reports to Congress on the Military Power of the People's Republic of China since 2000.

37. Philip J. Crowley, Assistant Secretary of State for Public Affairs, Daily Press Briefing, November 2, 2010.

38. Victoria Nuland, Press Secretary of the State Department, Daily Press Briefing, August 14, 2012.

39. Japan, *Law on the Exclusive Economic Zone and the Continental Shelf* (Law No. 74 of 1996), available from *www.un.org/Depts/los/LEGISLATIONANDTREATIES/PDFFILES/JPN_1996_Law74.pdf*.

40. China, *Exclusive Economic Zone and Continental Shelf Act*, June 26, 1998, available from *www.un.org/Depts/los/LEGISLATIONANDTREATIES/PDFFILES/chn_1998_eez_act.pdf*.

41. See the literature on China-Japan conflict over their efforts to explore fossil resources in the disputed area.

42. See the first UN-sponsored survey and report by K. O. Emery, Yoshikazu Hayashi, and others, "Geological Structure and Some Water Characteristics of the East China Sea and the Yellow Sea," UNECAFE/CCOP *Technical Bulletin*, No. 2, 1969.

43. Victor Prescott and Clive Schofield, *The Maritime Political Boundaries of the World*, 2nd Ed., Boston, MA: M. Nijhoff, 2005.

44. There is a vast literature on the Senkaku/Diaoyu dispute and the disputed maritime delimitation between China and Japan. China's recommended reading is by Zhong Yan, "On the

Sovereign Ownership of Diaoyu Dao," *People's Daily*, October 18, 1996, available from *www.xinhuanet.com.* (*news.xinhuanet.com/ziliao/2004-03/26/content_1386025.htm*). Japan's official stand is provided by the Ministry of Foreign Affairs of Japan, "The Basic View on the Sovereignty over the Senkaku Islands," "Q&A on the Senkaku Islands," and "Recent Japan-China Relations (October 2010)," available from *www.mofa.go.jp/region/asia-paci/senkaku/senkaku.html.* The best U.S. public document is the "Okinawa Reversion Treaty," Hearings before the Committee on Foreign Relations, United States Senate, 92 Congress, First Sess., Washington DC: U.S. Government Printing Office, October 27-29, 1971.

45. The key U.S. document is "Civil Administration Proclamation No. 27" about the geographical boundaries of the Ryukyu Islands by the U.S. Civil Administration of the Ryukyu Islands, Office of the Deputy Governor, APO 719, December 25, 1953.

46. Gao Jianjun, "China-Japan East China Sea Delimitation in the Context of the New International Law of the Sea," *Pacific Journal*, No. 8, 2005.

47. China first introduced its policy of "shelving disputes" on these islands in 1972 by then Chinese Prime Minister Zhou En-lai during the negotiation for resumption of diplomatic relations between China and Japan. China's second take on this policy was by then Vice Prime Minister Deng Xiaoping during his visit to Japan in 1978.

48. Ministry of Foreign Affairs of Japan, "Press Conference by Minister for Foreign Affairs Seiji Maehara," October 15, 2010, available from *www.mofa.go.jp/announce/fm_press/2010/10/1015_01.html.*

49. U.S. Department of State, "Briefing by Secretary Clinton, Japanese Foreign Minister Maehara," Honolulu, HI, October 27, 2010. Secretary of Defense Robert Gates and the Chairman of the Joint Chiefs of Staff Admiral Mike Mullen, when responding to Japanese NHK News interview questions, made the remarks that the United States would support its ally, Japan, and "We would fulfill our alliance responsibilities," quoted in Peter Lee, "High Stakes Gamble as Japan, China and the U.S. Spar in the East and South China Seas," *The Asia-Pacific Journal*, Vol. 43, October 25,

2010, pp. 1-10. Secretary of Defense Robert Gates also echoed the Secretary of State by saying that the United States would honor its military obligation in such a clash.

50. Ishihara is the author of *The Japan That Can Say No* [to the United States], New York: Simon & Schuster, 1991.

51. Ishihara has long wanted to purchase those islands, but the owner, Kunioki Kurihara, insisted that he would only sell the islands to the Japanese government, not any private individual. In this latest attempt, Ishihara would purchase the islands in the name of the Tokyo government. The Kurihara family agreed. See Masami Ito, "Owner OK with Metro Bid to Buy Disputed Senkaku Islands," *The Japan Times*, May 18, 2012.

52. Yoshitaka Unezawa, "Tokyo Governor Shintaro Ishihara Delivers His Speech in Washington on April 16, 2012," *The Asahi Shimbun AJW* (*Asia and Japan Watch*), April 18, 2012.

53. Ishihara quit his job as Governor of Tokyo in mid-October and formed a new party, "Taiyo no To" ("The Sunrise Party") in mid-November 2012. "With the establishment of the party, Ishihara aims to form a third force to take down the ruling Democratic Party of Japan and keep the Liberal Democratic Party, the main opposition force, from regaining power in the next election," Mizuho Aoki, *The Japan Times*, November 14, 2012.

54. *The Japan Times*, "Donations to Metro Government to Buy Senkaku Islands Top 1 Billion Japanese Yen," June 2, 2012.

55. Jiji, "Are Senkakus a 'Core Interest' for China?" *The Japan Times*, May 24, 2012.

56. The Japanese central government was reportedly to pay ¥2.05 billion (about $247 million) for the purchase. See *The Japan Times*, "Senkaku Purchase Bid Made Official," September 11, 2012.

57. Xu Chunliu, "China's Ocean Monitoring Command: Will Strengthen Control and Management on Disputed Maritime Territories," *New Capital Daily*, October 18, 2008, available from *military.people.com.cn/GB/8221/51755/141011/141012/8510843.html*.

58. Hai Tao, "Detailed Account of China's Maritime Surveillance Vessels to Diaoyu Dao: Japan Attempted to 'Hit' Chinese Vessels," *International Herald*, December 12, 2008.

59. *Xinhua net*, "Ministry of Foreign Affairs Spokesman: China Will Decide When to Send Monitoring Vessels to Diaoyu Dao," Press Conference, December 9, 2008.

60. Qi Lu, "China's Marine Surveillance Fleet Gets World Leading-Edge Vessels and Makes Regular Patrol in All China-Ruled Waters," *Ordinance*, February 27, 2009.

61. "Fishery Patrol Vessel Sets out for East China Sea," *China Daily*, November 16, 2010.

62. Zhang Yunbi, Wu Jiao, Tan Yingzi, and Fu Jing, "Tension Escalates as US Envoy Arrives," *China Daily*, October 17, 2012.

63. David S. Cloud, "On Way to Tokyo, Leon Panetta Urges Restraint in Islands Dispute," *Los Angeles Times*, September 16, 2012.

64. U.S. Department of State, "Media Roundtable in Tokyo, Japan," Interview with William J. Burns, Deputy Secretary, U.S. Embassy, Tokyo, Japan, October 15, 2012.

65. Nanae Kurashige, "Diplomat: U.S. Cannot Be Neutral over Senkaku," *The Asahi Shimbun-AJW*, October 31, 2012.

66. Nicholas J. Spykman, *The Geography of Peace*, New York: Harcourt, Brace and Co., 1994.

67. There is a huge literature on the 1997 Asian financial crisis and China's role in the crisis.

68. Andrew Walker, "China and ASEAN Free Trade Deal Begins," *BBC News*, January 1, 2010.

69. Lee Kuan Yew, "Speech by Minister Mentor Lee Kuan Yew at the U.S.-ASEAN Business Council's 25th Anniversary Gala Dinner in Washington DC," October 27, 2009.

70. Evelyn Goh, "Great Powers and Southeast Asian Regional Security Strategies: Omni-enmeshment, Balancing and Hierarchical Order," Singapore: Institute of Defense and Strategic Studies, 2005.

71. Han Zhenhua, *A Study of the History and Geography of the South China Sea Islands.* Beijing, China: Social Science Literature Publishing, 1996; Liu Nanwei, *A Documentation of the Names of the South China Sea.* Beijing, China: Science Publishing, 1996; "Spratly," "Spratas," "Paracel," "Macclesfield," and so on are European assignments and colonial legacies. See also Chinese Foreign Ministry, "The Issue of South China Sea Islands Sovereignty, [5] International Recognition of China's Sovereignty over the Spratly Islands"), available from *www.mfa.gov.cn/chn/pds/ziliao/tytj/t10651.htm.*

72. See Sun Donghu, "The Origins of South China Sea islands' Foreign Names and Their Historical Influence," *Geography Studies*, No. 2, 2000.

73. The Soviet Union proposed an amendment to Article 2 (b) and (f) as follows: "Japan recognizes full sovereignty of the Chinese PRC over Manchuria, the Island of Taiwan (Formosa) with all islands adjacent to it, the Penlinletao Islands (Pescadores), the Tunshatsuntao (Pratas Islands), as well as over the islands of Sishatsuntao and Chunshatsuntao (the Paracel Islands), and Nanshatsuntao Islands, including the Spratlys, and renounces all right, title, and claim to the territories named herein," The delegates rejected the Soviet amendment by a vote of 46 to 3, with 1 abstention. See U.S. State Department, *Conference for the Conclusion and Signature of the Peace Treaty with Japan, Record of Proceedings*, Washington, DC: U.S. Department of State, Publication 4392, December 1951, p. 292.

74. The Prime Minister of the newly established Republic of Vietnam, Tran Van Huu, who attended the peace conference at San Francisco, claimed Vietnam's ownership of the South China Sea islands during his speech at the conference (every delegate had at least one opportunity to speak at the conference, the big powers, of course, got more than their share of the time to speak there). However, there was no action taken on that claim—it went unnoticed. See U.S. State Department, *Conference for the Conclusion*

and Signature of the Peace Treaty with Japan, Record of Proceedings.
Washington, DC: U.S. Department of State, Publication 4392, December 1951, p. 292. The Vietnam White Paper of 1975 claims that since there was no objection to the Prime Minister's claim at the conference, it was endorsed. This interpretation cannot stand. See also Li Jinming, "Vietnam's Claim on the South China Sea Territory," *View on Southeast Asia* (China), No. 1, 2005, for an analysis of the issues.

75. A draft was circulated prior to the conference. China also obtained a copy.

76. Statement by Chinese Foreign Minister Zhou En-lai in *Collected Documents on the Foreign Relations of the People's Republic of China.* Beijing, China: World Knowledge Publishing, 1961.

77. See Chinese Captain Memoir and Vietnam White Paper.

78. International practice gives more weight to effective control than historical claims. See the Island of Palmas Case ruled by the Permanent Court of Arbitration in 1925 (a dispute between the United States on behalf of its colony the Philippines and the Netherlands on behalf of its colony Indonesia); it sets the example of effective exercise of sovereignty taking precedence over the right of discovery and claim of territorial contiguity (see reference Hague Court Reports 2d 83 [1932] [Perm. Ct. 4rb. 1928]). Other cases such as the Clipperton Island, Eastern Greenland, and the Minqulers and Ecrehos Islands all point to the same requirement of effective control over historical claims.

79. The Republic of China (ROC) did station some troops in the Paracel and Spratly Islands. However, following its defeat in the Chinese Civil War, it withdrew the troops. The PRC did not replace them with PLA forces, leaving those islands largely unattended.

80. Chinese Government (PRC) Declaration on China's Territorial Waters, September 4, 1958, available from *news3.xinhuanet. com/ailiao/2003-01/24/content_705061.htm.* "Historically belonging to China" and "China's intrinsic and inseparable territories" are two dogmatic terms in China's claims to its disputed territories. China and Vietnam share borders. China also owned and ruled Vietnam for well over 1,000 years. China can claim its historical

ties and inseparable connections with Vietnam and demand its return to China, saying, "Vietnam historically belongs to China; it is an inseparable part of China." International practice (International Court and Arbitration) does not accept congruity as a reason for territorial claim.

81. Chinese documentation of activities by all relevant parties, including the Europeans, throughout the ages in the South China Sea is available from *www.nansha.org.cn*.

82. Liselotte Odgaard, *Maritime Security between China and Southeast Asia: Conflict and Cooperation in the Making of Regional Order*, Surrey, UK: Ashgate, 2002, pp. 67-76.

83. Due to definitional differences, the number of islands in the Spratlys is different, and so is the number of islands each disputant holds. See John C. Baker and David G. Wiencek, eds., *Cooperative Monitoring in the South China Sea*. Westport, CT: Praeger Publishers, 2002; Dieter Heinzig, *Disputed Islands in the South China Sea*, Hamburg, Germany: Hamburg Institute of Asian Affairs, 1976; Marwyn S. Samuels, *Contest for the South China Sea*. New York: Metheun, 1982; Li Jinming, "The Current Status of the South China Sea Disputes," *Southeast Asian Studies* (China), No. 1, 2002; Wikipedia, "Spratly Islands"; Lu Ning, *Flashpoint Spratlys*, Singapore: Dolphin Trade Press, 1995.

84. Cai Penghong, "An Analysis of the U.S. South China Sea Policy," *Contemporary International Relations* (China), No. 9, 2009; Guo Yuan, "U.S. South China Sea Policy in the Post-Cold War Era," *Academic Inquiry* (China), February 2008; He Zhigong and An Xiaoping, "The U.S. Factor in the South China Sea Disputes," *Journal of Contemporary Asia-Pacific Studies* (China), No. 1, 2010; Liu Zhongmin, "An Analysis of the U.S.-China Relationship and Its Impact on the South China Sea Issues," *Northeast Asia Forum* (China), Vol. 15, No. 5, September 2006; Liu Zhongmin, "South China Sea Issues and Post-Cold War U.S.-China Relations: Conflicting Understanding and Difficult Choices," *Foreign Affairs Review* (China), Issue 85, December 2005; Qiu Danyang, "The U.S. Factor in Sino-Filipino South China Sea Dispute," *Journal of Contemporary Asia-Pacific Studies* (China), No. 5, 2002.

85. Hu Suping, "The Evolution of U.S. Policy on the South China Sea, 1950-2004," *The New Orient* (China), No. 5, 2010. Hu mentions in her article that the Chinese Foreign Ministry had issued more than 400 protests against the United States between September 1956 and 1970. But Hu provides no source or reference for this number (a common problem with most Chinese publications).

86. Joseph Nye, Jr., *Far Eastern Economic Review*, 1995, p. 22.

87. Warren Christopher, quoted in A. James Gregor, "Qualified Engagement: U.S. China Policy and Security Concerns," *Naval War College Review*, Spring 1999. Gregor states in his article that "U.S. Secretary of State Warren Christopher reminded the Chinese foreign minister that the United States had treaty obligations with the Philippines," The reference of this remark is from *Indochina Digest*, April 21, 1995.

88. An influential piece on the U.S. neutrality on the South China Sea disputes is by Scott Snyder, "The South China Sea Dispute: Prospects for Preventive Diplomacy," Special Report to the United States Peace Institute, 1996.

89. Richard Cronin, "Maritime Territorial Disputes and Sovereignty Issues in Asia," A Testimony before the Senate Subcommittee on East Asian and Pacific Affairs, July 13, 2009; Peter A. Dutton, "China's Views of Sovereignty and Methods of Access Control," Testimony before the U.S.-China Economic and Security Review Commission, February 27, 2008; Dan Blumenthal, "Hearing on Maritime Territorial Disputes in East Asia," Testimony before the Senate Foreign Relations Committee Subcommittee on Asia, July 15, 2009.

90. Gloria Jane Baylon, "U.S. Should Support the Republic of the Philippines' Claim on Spratlys," *Philippines News Agency* (PNA), March 4, 2009; U.S. Senator Jim Webb, "The Implications of China's Naval Modernization for the United States," Testimony before the U.S.-China Economic and Security Review Commission, June 11, 2009; Federal News Service, Hearing of the East Asian and Pacific Affairs Subcommittee of the Senate Foreign Relations Committee Subject: Maritime Disputes and Sovereignty Issues in East Asia, Chaired by Senator Jim Webb (D-VA), July

15, 2009; Federal Information and News Dispatch, Inc., "Senator Webb Completes Landmark Five Nation Visit throughout Asia," August 24, 2009; Greg Torode, "A Classic Display of McCain Resilience," *South China Morning Post*, April 11, 2009; Jason Folkmanis, "U.S. Vietnam Seek to Limit China, Keep Power Balance," *Bloomberg News*, August 28, 2009, available from *www.bloomberg.com/apps/news?pid=newsarchive&sid=a1ax5sgs8b0Q*.

91. Hillary Rodham Clinton, Secretary of State, "Remarks with Thai Deputy Prime Minister Korbsak Sabhavasu," Government House, Bangkok, Thailand, July 21, 2009; Hillary Rodham Clinton, Secretary of State, "Remarks at the ASEAN Regional Forum," National Convention Center, Hanoi, Vietnam, July 23, 2010, available from *www.state.gov/secretary/rm/2010/07/145095.htm*.

92. Robert Gates, "Remarks at the 9th IISS Asia Security Summit, the Shangri-La Dialogue," Singapore, June 5, 2010.

93. Office of the Press Secretary, "Remarks by President Barack Obama at Suntory Hall, Tokyo, Japan," Washington, DC: The White House, November 14, 2009.

94. *Ibid.*

95. Wu Xingtang, "The Fluctuating Nature of U.S.-China Relationship and Its Difficult Journey," *The Red Flag Journal*, (China), No. 20, 2010; Zhao Minghao, "'Return' or 'Repositioning': an Analysis of the U.S. Adjustment of Its Strategy toward the Asia-Pacific," *Contemporary World* (China), No. 12, 2010.

96. See one interesting reflection from Su Hao, Director of the Center for Strategic and Conflict Management at China Foreign Affairs University in Beijing, "Washington's High Ambition for an East Asian Presence," *China Daily*, August 9, 2010.

97. Edward Wong, "China Hedges over Whether South China Sea Is a 'Core Interest' Worth War," *The New York Times*, March 30, 2010. Steinberg and Bader also shared this new development with the policy circle on various occasions.

98. Bonnie Glaser and David Szerlip, "U.S.-China Relations: The Honeymoon Ends," *Comparative Connections*, April 2010.

99. Remarks by the Secretary of State Clinton at the ASEAN Regional Forum, National Convention Center, Hanoi, Vietnam.

100. The 27 foreign ministers are from 10 ASEAN nations: Brunei, Cambodia, Indonesia, Laos, Malaysia, Myanmar, the Philippines, Singapore, Thailand, and Vietnam; 10 ASEAN dialogue partners: Australia, Canada, China, the European Union, India, Japan, New Zealand, South Korea, Russia, and the United States; and seven Asia-Pacific nations: Bangladesh, North Korea, Pakistan, Mongolia, Sri Lanka, Timor-Leste, and Papua New Guinea.

101. The Monroe Doctrine is articulated in a message by President James Monroe to the Congress on December 2, 1823. It asserted that the Western Hemisphere was not to be further colonized by European countries, but that the United States would neither interfere with existing European colonies nor meddle in the internal concerns of European countries. The Carter Doctrine came in the aftermath of the Soviet invasion of Afghanistan in 1980. There was concern that the Soviets may continue its advance down to the Persian Gulf, threatening to take control of the world's oil reserve. President Jimmy Carter made the volatile Middle East and Central Asia a focal point of his State of the Union address on January 23, 1980. In this speech, Carter said, "Let our position be absolutely clear: An attempt by any outside force to gain control of the Persian Gulf region will be regarded as an assault on the vital interests of the United States of America, and such an assault will be repelled by any means necessary, including military force." President Carter's message is contained in the *Weekly Compilation of President Documents*, Vol. XVI, January 28, 1980, pp. 194-200. See also Cecil V. Crabb, Jr., *The Doctrines of American Foreign Policy*, Baton Rouge, LA: Louisiana State University Press, 1982, for an excellent discussion of U.S. foreign policy doctrines.

102. *Reuters*, "China on the Defensive at ARF over Maritime Rows," Hanoi, Vietnam, July 24, 2010. See the Chinese Foreign Minister's response, "Foreign Minister Yang Jiechi Refutes Fallacies on the South China Sea Issue" at the Chinese Foreign Ministry Website, available from *www.mfa.gov.cn/eng/zxxx/t719460. htm*. Also see Andrew Jacobs, "China Warns U.S. to Stay out of Islands Dispute," *The New York Times*, July 26, 2010; and Jay Solomon, "U.S. Takes on Maritime Spats," *The Wall Street Journal*, July 24, 2010.

103. Liu Fengan, "Our Military's Busy Exercises in 2010, Large-scale and Joint Exercises Have Become Regular Practices," *Outlook Oriental Weekly* (China), December 13, 2010.

104. Li Jinming, "South China Sea Problem: The United States Moving from Neutrality to High-Profile Involvement," *World Knowledge* (China), Issue 24, 2010; Cheng Hanping, "An Analysis of the Policy and Practice of U.S. Open Involvement on South China Sea Disputes," *Southeast Asia Forum* (China), No. 2, 2010.

105. Xie Xiaojun, "What Is the U.S. Intent in the South China Sea Affairs?" *Current Affairs Analysis* (China), 2010.

106. Jeff Mason and James Pomfret, "Obama Urges Restraint in Tense Asian Disputes," *Reuters*, November 20, 2012.

107. Office of the Press Secretary, "Fact Sheet: East Asia Summit Outcomes," Washington, DC: The White House, November 20, 2012.

APPENDIX 1

CAIRO DECLARATION

Conference of President Roosevelt, Generalissimo Chiang Kai-shek, and Prime Minister Churchill in North Africa.[1]

President Roosevelt, Generalissimo Chiang Kai-shek, and Prime Minister Churchill, together with their respective military and diplomatic advisers, have completed a conference in North Africa.

The following general statement was issued:

"The several military missions have agreed upon future military operations against Japan. The Three Great Allies expressed their resolve to bring unrelenting pressure against their brutal enemies by sea, land, and air. This pressure is already rising.

"The Three Great Allies are fighting this war to restrain and punish the aggression of Japan. They covet no gain for themselves and have no thought of territorial expansion. It is their purpose that **Japan shall be stripped of all the islands in the Pacific which she has seized or occupied since the beginning of the First World War in 1914, and that all the territories Japan has stolen from the Chinese, such as Manchuria, Formosa, and the Pescadores, shall be restored to the Republic of China.** Japan will also be expelled from all other territories which she has taken by violence and greed. The aforesaid three great powers, mindful of the enslavement of the people of Korea, are determined that in due course Korea shall become free and independent.

"With these objects in view that three Allies, in harmony with those of the United Nations at war with Japan, will continue to persevere in the serious and prolonged operations necessary to procure the unconditional surrender of Japan."

ENDNOTE - APPENDIX 1

1. Released to the Press by the White House, December 1, 1943. Source: The Department of State Bulletin, Vol. IX, No. 232, Washington DC, December 4, 1943 (bold face emphasis added).

APPENDIX 2

POTSDAM PROCLAMATION

Defining Terms for Japanese Surrender.[1]

(1) We—the President of the United States, the President of the National Government of the Republic of China, and the Prime Minister of Great Britain, representing the hundreds of millions of our countrymen, have conferred and agree that Japan shall be given an opportunity to end this war.

(2) The prodigious land, sea, and air forces of the United States, the British Empire, and of China, many times reinforced by their armies and air fleets from the west, are poised to strike the final blows upon Japan. This military power is sustained and inspired by the determination of all the Allied Nations to prosecute the war against Japan until she ceases to resist.

(3) The result of the futile and senseless German resistance to the might of the aroused free peoples of the world stands forth in awful clarity as an example to the people of Japan. The might that now converges on Japan is immeasurably greater than that which, when applied to the resisting Nazis, necessarily laid waste to the lands, the industry and the method of life of the whole German people. The full application of our military power, backed by our resolve, *will* mean the inevitable and complete destruction of the Japanese armed forces and just as inevitably the utter devastation of the Japanese homeland.

(4) The time has come for Japan to decide whether she will continue to be controlled by those self-willed militaristic advisers whose unintelligent calculations have brought the Empire of Japan to the threshold of annihilation, or whether she will follow the path of reason.

(5) Following are our terms. We will not deviate from them. There are no alternatives. We shall brook no delay.

(6) There must be eliminated for all time the authority and influence of those who have deceived and misled the people of Japan into embarking on world conquest, for we insist that a new order of peace, security, and justice will be impossible until irresponsible militarism is driven from the world.

(7) Until such a new order is established *and* until there is convincing proof that Japan's war-making power is destroyed, points in Japanese territory to be designated by the Allies shall be occupied to secure the achievement of the basic objectives we are here setting forth.

(8) **The terms of the Cairo Declaration shall be carried out and Japanese sovereignty shall be limited to the islands of Honshu, Hokkaido, Kyushu, Shikoku and such minor islands as we determine.**

(9) The Japanese military forces, after being completely disarmed, shall be permitted to return to their homes with the opportunity to lead peaceful and productive lives.

(10) We do not intend that the Japanese shall be enslaved as a race or destroyed as a nation, but stern justice shall be meted out to all war criminals, including those who have visited cruelties upon our prisoners. The Japanese

Government shall remove all obstacles to the revival and strengthening of democratic tendencies among the Japanese people. Freedom of speech, of religion, and of thought, as well as respect for the fundamental human rights, shall be established.

(11) Japan shall be permitted to maintain such industries as will sustain her economy and permit the exaction of just reparations in kind, but not those which would enable her to re-arm for war. To this end, access to, as distinguished from control of, raw materials shall be permitted. Eventual Japanese participation in world trade relations shall be permitted.

(12) The occupying forces of the Allies shall be withdrawn from Japan as soon as these objectives have been accomplished and there has been established in accordance with the freely expressed will of the Japanese people a peacefully inclined and responsible government.

(13) We call upon the government of Japan to proclaim now the unconditional surrender of all Japanese armed forces, and to provide proper and adequate assurance of their good faith in such action. The alternative for Japan is prompt and utter destruction.

ENDNOTE - APPENDIX 2

1. This proclamation issued on July 26, 1945, by the heads of governments of the United States, the United Kingdom, and China. It was signed by the President of the United States and the Prime Minister of the United Kingdom at Potsdam and concurred with by the President of the National Government of China, who communicated with President Truman by dispatch. Source: The Department of State Bulletin, Vol. XIII, No. 318, Washington DC, July 29, 1945 (bold face emphases added).

TREATY OF PEACE WITH JAPAN[1]

Chapter II

Territory

Article 2

(a) Japan, recognizing the independence of Korea, renounces all right, title, and claim to Korea, including the islands of Quelpart, Port Hamilton, and Dagelet.

(b) **Japan renounces all right, title, and claim to Formosa and the Pescadores.**

(c) Japan renounces all right, title, and claim to the Kurile Islands, and to that portion of Sakhalin and the islands adjacent to it over which Japan acquired sovereignty as a consequence of the Treaty of Portsmouth of 5 September 1905.

(d) Japan renounces all right, title, and claim in connection with the League of Nations Mandate System, and accepts the action of the United Nations Security Council of 2 April 1947, extending the trusteeship system to the Pacific Islands formerly under mandate to Japan.

(e) Japan renounces all claims to any right or title to or interest in connection with any part of the Antarctic area, whether deriving from the activities of Japanese national or otherwise.

(f) **Japan renounces all right, title, and claim to the Spratly Islands and to the Paracel Islands.**

ENDNOTE - APPENDIX 3

1. Neither the Republic of China (ROC) in Taiwan nor the People's Republic of China (PRC) in mainland China were invited because of the Chinese Civil War and the controversy over which government was a legitimate representative of China. A total of 51 nations attended the conference, but **48 nations signed** at San Francisco on September 8, 1951; the Soviet Union, Czechoslovakia, and Poland refused to do so. Source: United Nations Treaty Series 1952 (reg. no. 1832), Vol. 136, pp. 45-164 (bold face emphases added).